Computational and Instrumental Methods in EPR

Volume 25

Christopher J. Bender and
Lawrence J. Berliner

Computational and Instrumental Methods in EPR

Volume 25

 Springer

Christopher J. Bender
Department of Chemistry
Fordham University
441 East Fordham Road
Bronx, NY 10458
USA
bender@fordham.edu

Lawrence J. Berliner
Department of Chemistry &
Biochemistry
University of Denver
2190 E. Iliff Avenue
F. W. Olin Hall, Room 202
Denver, CO 80208
USA
berliner@du.edu

ISBN 978-1-4419-4112-1

e-ISBN-10: 0-387-38880-X
e-ISBN-13: 978-0387-38809

Printed on acid-free paper.

9 8 7 6 5 4 3 2 1

springer.com

dedicated to

Arthur Schweiger (1946–2006)

*the acknowledged master of spin acrobatics,
pulsed EPR/ESR experiments, and methodology*

CONTRIBUTORS

Christopher J. Bender
Department of Chemistry
Fordham University
441 East Fordham Road
Bronx, New York 10458, USA

Robert Lopez
Laboratoire de Magnétisme et d'Electronique Quantique
Université Paul Sabatier Toulouse III
31077 Toulouse CEDEX, France

Sushil K. Misra
Physics Department
Concordia University
1455 de Maisonneuve Boulevard West
Montreal, Québec H3G 1M8, Canada

Y. Shimoyama
Department of Physics
Hokkaido University of Education
Hakodata, 040-0083 Japan

H. Watari
Uchihonmachi 2-chome
Suita, Osaka, 564-0032 Japan

PREFACE

Electron magnetic resonance in the time domain has been greatly facilitated by the introduction of novel resonance structures and better computational tools, such as the increasingly widespread use of density-matrix formalism. This second volume in our series, devoted both to instrumentation and computation, addresses applications and advances in the analysis of spin relaxation time measurements.

Chapters 1 deals with the important problem of measuring spin relaxation times over a broad temporal range. The author, Dr. Sushil Misra, has worked on a wide variety of solutions to problems in this area, with respect to both experimental and theoretical aspects, and Chapter 1 summarizes much of his recent work, which was enhanced by a fruitful collaboration with the late Professor Jacques Pescia. Chapter 2 presents solutions to the problem of measuring short spin relaxation times. Again, in collaboration and tribute to the late Jacques Pescia's laboratory, part of the chapter represents a translation of the amplitude modulation technique section from a doctoral thesis by Robert Lopez in 1993 to The Paul Sabatier University. Experimental data that appeared in the original thesis are placed at the end of subsections that correspond to the described technique.

Chapter 3 takes up the problem of multi-frequency ENDOR and ESEEM, and illustrates how small stepwise increments of spectrometer operating parameters can enable one to better determine spin-Hamiltonian parameters via a graphical analysis.

Chapters 4 and 5 address computational problems of EMR, particularly the treatment of high spin systems. Chapter 4, describes details of simulating Mn(II) EPR spectra in single crystals, polycrystalline, and amorphous materials. The analytical technique is based on eigenvalues of the spin-Hamiltonian matrix calculated to third order in perturbation for orthorhombic distortion, so as to estimate values of the zero-field splitting parameters — D, E — from forbidden hyperfine doublet separations in the central sextet of an Mn(II) EPR spectrum. Also included is a quick, precise simulation of EPR spectra in amorphous materials by matrix diagonalization using the method of homotopy, wherein the resonant field values are obtained by the method of least-squares fitting from the knowledge of their values at an infinitesimally close orientation of the external magnetic field. Several examples of computed spectra in amorphous materials are given and characteristics of amorphous spectra are discussed.

Finally, Chapter 5 by Hiroshi Watari and Yuhei Shimoyama describes density-matrix formalism for treating angular momenta in multi-quantum systems. The spectrum resolution theorem is used to obtain a linear combination representation of the spin Hamiltonian and greatly simplifies the manipulation of angular momenta with high quantum numbers.

C. J. Bender
New York, New York

L. J. Berliner
Denver, Colorado

CONTENTS

Chapter 3

Quantitative Measurement of Magnetic Hyperfine Parameters and the Physical Organic Chemistry of Supramolecular Systems

Christopher J. Bender

Chapter 4

New Methods of Simulation of Mn(II) EPR Spectra: Single Crystals, Polycrystalline and Amorphous (Biological) Materials

Sushil K. Misra

Chapter 5

**Density Matrix Formalism of Angular Momentum in
Multi-Quantum Magnetic Resonance**

H. Watari and Y. Shimoyama

MICROWAVE AMPLITUDE MODULATION TECHNIQUE TO MEASURE SPIN–LATTICE (T_1) AND SPIN–SPIN (T_2) RELAXATION TIMES

Sushil K. Misra

Physics Department, Concordia University, 1455 de Maisonneuve Boulevard West, Montreal Quebec H3G 1M8 Canada

1. INTRODUCTION

The measurement of very short spin–lattice, or longitudinal, relaxation (SLR) times (*i.e.*, $10^{-10} < T_1 < 10^{-6}$ s) is of great importance today for the study of relaxation processes. Recent case studies include, for example, glasses doped with paramagnetic ions (Vergnoux *et al.*, 1996; Zinsou *et al.*, 1996), amorphous Si (dangling bonds) and copper–chromium–tin spinel (Cr^{3+}) (Misra, 1998), and polymer resins doped with rare-earth ions (Pescia *et al.*, 1999a; Pescia *et al.* 1999b). The ability to measure such fast SLR data on amorphous Si and copper-chromium-tin spinel led to an understanding of the role of exchange interaction in affecting spin–lattice relaxation, while the data on polymer resins doped with rare-earth ions provided evidence of spin–fracton relaxation (Pescia *et al.*, 1999a,b). But such fast SLR times are not measurable by the most commonly used techniques of saturation- and inversion-recovery (Poole, 1982; Alger, 1968), which only measure spin–lattice relaxation times longer than 10^{-6} s. A summary of relevant experimental data is presented in Table 1.

It is possible to monitor relaxation times by modulating the microwave amplitude at a rate close to the inverse spin-lattice relaxation time, T_1^{-1} (Misra, 2004). This is due to the fact that if the microwave amplitude is changed faster than that which the spins can follow as governed by the electron SLR time, the EPR signal is determined predominantly by T_1. This modulation technique permits measurement of relaxation times in the interval between the values that are slow enough to be measured by the recovery techniques and the values that are too fast to have any significant impact on continuous wave (cw) lineshape. These latter relaxation times can be as short as 10^{-10} s, which is substantially shorter than those that can be measured by the saturation/inversion recovery techniques.

Table 1. Spin–Lattice Relaxation Rates Due to Exchange Interaction For Various Amorphous Mmaterials at Selected Temperatures (Misra, 1998).

	$1/T_1$ (s^{-1}) at temperature indicated				
	10 K	20 K	50 K	100 K	200 K
Amorphous silicon (Gourdon et al., 1981)	10^5	5×10^5	2×10^6	10^7	10^8
Amorphous silicon (Askew et al., 1986)		5×10^4 (~100 at 1 K)			
Borate glass, 0.1% Fe_2O_3 (Zinsou et al., 1996)		$\sim10^6$	$\sim7\times10^6$	2×10^7	4×10^7
Borate glass, 0.5% Fe_2O_3 (Zinsou et al., 1996)		$\sim10^7$	3×10^7	4.5×10^7	5×10^7
MgO:P_2O_5, 0.2% Mn (Vergnoux et al., 1998)					2.5×10^6
$Cu_{2x}Cr_{2x}Sn_{2-2x}S_4$ (x = 0.8) (Sarda et al., 1989)	5×10^8	8×10^8	10^9	5×10^9	4×10^9

There have been extensive studies of fast relaxation time measurements using the amplitude modulation technique (Hervé & Pescia, 1960, 1963a,b; Pescia & Hervé, 1963; Ablart & Pescia, 1980; Ablart, 1978). In each of these studies the EPR signal along the z-axis (defined to be the direction of external magnetic field B_0), which is proportional to magnetic moment M_z, was monitored by an electromotic force (emf) that is induced in a pickup coil. These original measurements, however, suffered from limitations, both instrumental and computational. As concerns the instrument and the collection of data, the introduction of the pickup coil has a deleterious effect upon the signal-to-noise (S/N) ratio of the spectrometer and its associated sample resonator. And as regards the analysis of the data, the solutions to the relevant differential equations require that one use asymptotic methods (as opposed to an exact solution) and limit the experimental condition to small amplitude modulation depths (Pescia, 1965).

The measurement of fast relaxation times via the amplitude modulation technique may be improved by recent technological advances (G. Eaton & S. Eaton, personal communication). These technologies include: (i) modern lock-in amplifiers that operate at high frequency (*e.g.*, 200 MHz, Stanford Research SR844) used together with (ii) a double-balanced mixer to provide high-speed modulation of the microwave amplitude, and (iii) the crossed-loop resonator (CLR) (Rinard *et al.*, 1994, 1996a,b). The CLR permits detection of the EPR signal in the *x,y*-plane (perpendicular to the direction of the external magnetic field), avoiding the need to use a pickup coil resonated at a single modulation frequency, which suffers from some serious drawbacks, as used in the Hervé and Pescia spectrometer. Moreover, the CLR permits one to use larger depths of amplitude, with a resultant enhancement in the signal-to-noise ratio.

The purpose of this chapter is to review the amplitude modulation technique, address the limitations of the classic studies in the light of these recent technological advances, and present equations that can be used to make predictions about both the EPR response in the *z*- and *x,y*-directions.

And because the technology and desired data are optimized by using large modulation depths, the limitations of data analysis will be redressed by describing a quick matrix technique for the solution of Bloch's phenomenological equations for the case when the amplitude of the microwave field is sinusoidally modulated with an arbitrary coefficient of modulation. The resultant solutions can then be used to calculate the signal obtained by either a classic pickup coil or cw-EPR signal in a resonator. With these solutions to the Bloch equations and analysis of detected signals, one may evaluate relaxation times T_1 and T_2 accurately by the use of a rigorous least-squares fitting procedure.

2. DESCRIPTION OF THE MICROWAVE AMPLITUDE MODULATION TECHNIQUE

As introduced, the method of measuring spin relaxation times via amplitude modulation employed a pickup coil for detecting the magnetization M_z and a double modulation scheme in order to accommodate the relatively low operating frequencies (*ca.* 1 kHz) of lock-in amplifiers at that time (Hervé & Pescia, 1960; Hervé & Pescia, 1963a; Hervé & Pescia, 1963b; Ablart & Pescia, 1980; Ablart, 1978; Pescia, 1965). In practice, two orthogonal fields $\mathbf{B}_0$ (external magnetic field) and $\mathbf{B}_1$ (microwave magnetic field) are applied to the sample. The amplitude of $\mathbf{B}_1$ is modulated at the frequency $\Omega/2\pi$ (this frequency is on the order of 1-10 MHz). This causes M_z, the component of the magnetization along $\mathbf{B}_0$, in a cw-EPR spectrometer to become modulated, which, in turn, induces an emf in a pickup coil placed close to the sample and having its axis directed along the external magnetic field. The induced signal is proportional to dM_z/dt and is detected by a lock-in amplifier.

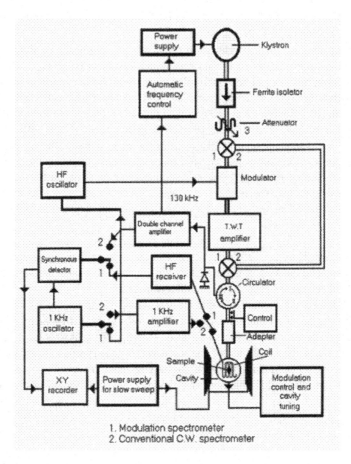

1. Modulation spectrometer
2. Conventional C.W. spectrometer

Figure 1. Block diagram of a spectrometer for conventional (connections 1) and amplitude modulation (connections 2) EPR (after Vergnoux, 1996).

The modulation spectrometer is depicted as a block diagram in Figure 1 (Ver-
genoux, 1996; Ablart & Pescia, 1980; Ablart, 1978). A resonant continuous-wave
signal is amplitude modulated by a high-frequency (HF) generator at frequency
$\Omega/2\pi$. (The spectrometer is operable in bands at 0.2, 0.7, 4, or 8–12 GHz so as to
observe the effect of $\mathbf{B}_0$ on the relaxation time (*cf.* Ablart & Pescia, 1980).) This
high-frequency AM signal is, in turn, amplitude modulated at 1 kHz by a second
(IF) oscillator in order to permit lock-in detection of the EPR signal. In other
words, the high-frequency modulation causes $|M_z|$ to oscillate within range ΔM_z
over period $(\Omega/2\pi)^{-1}$; the 1-kHz signal that is then imposed on the $\Omega/2\pi$ signal (for
the purpose of lock-in detection) serves to vary the range (*i.e.*, $\Delta M_z = M_{z,\text{HI}} - M_{z,\text{LO}}$)
over a 1-ms time 'frame'.

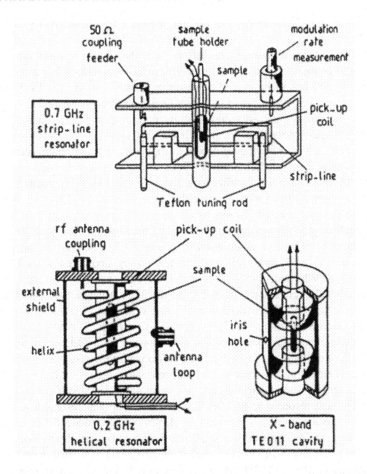

Figure 2. Details of resonator contruction. Pick-up coils are wound around the sample holder tube, tuned to frequency $\Omega/2$, and connected to a selective receiver via a low-noise impedance matching circuit (after Ablart & Pescia, 1980).

As described by Pescia and coworkers, pickup coils were devised to operate with sample resonators that operated at specific microwave frequencies, that is, a wire helix (0.2 GHz), a stripline (0.7 GHz), and a cavity (4, 8–12 GHz, operated in reflection mode). As a representative example, the cavity resonator that is used at X-band is illustrated in Figure 2 (detailed diagrams of the helix and stripline resonators may be found in Ablart & Pescia, 1980). Here the saddle-shaped pickup coil is wound around the sample holder tube inside the cavity. In each case the pickup coil is tuned at frequency $\Omega/2\pi$ by a variable capacitor and connected to a selective receiver via a low-noise impedance matching circuit.

But despite many attempts to optimize the use of a pickup coil (*e.g.*, use of matched bucking coils to cancel noise), the AM measurement technique as described in the preceding paragraphs suffers from the following disadvantages: (i) the S/N ratio is poor; (ii) many coil turns are needed to obtain a large induced voltage; (iii) the coil has to be placed outside the resonator where the EPR signal is small; (iv) the coil must be tuned to a high Q at a particular modulation frequency in order to get a large induced voltage; (v) the coil tends to pick up the modulation signal itself superimposed on the modulated EPR signal; (vi) any change in the magnetic field induces a voltage in the coil, imposing the requirement of having an extremely stable magnetic field; and (vi) coils tend to be microphonic.

It is therefore desirable to eliminate pickup coils from the detection scheme. With conventional EPR spectrometers, the normal mode of detection is via the microwave power reflected from a tuned cavity resonator and, in principle, this same reflection mode detection may be used to monitor the change in EPR signal as a function of microwave modulation frequency ($\Omega/2\pi$). But the desired signal is small and buried in the large reflected incident power modulated at the frequency Ω (Weidner & Whitmer, 1952). A bimodal sample resonator, however, supports two resonator modes that are orthogonal and therefore isolated. Separately used for sample excitation and signal detection, the two orthogonal modes enables one to isolate the microwave source (and its inherent noise) from the detected signal. The bimodal sample resonator therefore obviates the need for a pickup coil and avoids the problem of trying to extract a small signal from a large noise background.

The problem with constructing bimodal resonators, however, has been the fact that the magnetic and electric fields are never truly isolated from each other. Lumped element resonators, such as the so-called loop-gap resonator (LGR), are better suited to accomplish separation of electric and magnetic fields, at least to regions which are recognizable primarily as inductors or capacitors. The resultant crossed loop-gap resonator (Rinard *et al.*, 1996a,b (3 GHz); Rinard *et al.*, 2000 (L-band); Rinard *et al.*, 2002 (250 MHz)) is based on this idea, wherein one uses two resonators arranged orthogonal to one another, with only the sample region being in common. This is equivalent to a transmission between the resonators wherein the input and output are coupled only by the spin system. Other bimodal resonators are cited by Rinard *et al.* (2000). An alternative design in which parallel or antiparallel fields achieve isolation within an S-band loop-gap resonator by tuning so that the integral over all the space of the scalar products of the two modes is zero is provided by Piasecki *et al.*, 1996.

The importance of a crossed-loop or other bimodal resonator is that they offer the opportunity to measure the x,y-components of the spin magnetization. The incentive for the development of the theory of modulation spectroscopy in this chapter is (a) to extend it to deeper modulation than in prior treatments, and (b) predict the signals observable with x,y-detection as well as in the z-direction. Each may have its niche in measurements of spin relaxation times.

3. CALCULATION OF THE MODULATION SIGNAL BY SOLVING BLOCH'S EQUATIONS IN THE PRESENCE OF AMPLITUDE MODULATION

3.1. Definitions and the Impact of Modulation on the Experimental Arrangement for Detecting the Signal

We wish to compare the signal, S, obtained under two experimental scenarios, namely, that of the pickup coil vs. the crossed-loop resonator. We shall begin by deriving the equations that describe the signal induced in a pickup coil during an amplitude modulation experiment. In the following discussion, it will be given that the coil is tuned to modulation frequency $\Omega/2\pi$ and that the sample is small relative to the mean diameter of the coil. Under these preliminary assumptions, it can be stated that the induced signal in a pickup coil is (Ablart & Pescia, 1980; Ablart, 1978):

$$S_\alpha \text{ (pickup)} = nQ_b \frac{\mu_0}{2R} \frac{dM_\alpha}{dt} \text{ ; } (\alpha = z, y) \tag{1}$$

In eq. (1), Q_b is the filling factor; n is the number of turns in the coil; M_z and M_y are the z and y components of the magnetization, respectively; μ_0 is the permeability of the free space; and R is the coil radius. By comparison, the induced continuous-wave signals in a resonator are given by:

$$S_\alpha \text{(resonator)} \propto M_\alpha \text{ ; } (\alpha = z, y) \tag{2}$$

The amplitude-modulated microwave field is defined as (Hervé & Pescia, 1960, 1963a,b; Pescia & Hervé 1963; Ablart & Pescia, 1980; Ablart, 1978):

$$\vec{B}_1(t) = \vec{B}_1(1 + m_c \cos \Omega t)/(1 + m_c) \tag{3}$$

in which m_c is defined as the coefficient of modulation for a pickup coil and as equal to the fractional depth of the modulation. Here Ω corresponds to the modulation frequency, and in practice $m_c \ll 1$. In order to describe deeper modulation, one must revise eq. (3):

$$\vec{B}_1(t) = \vec{B}_1 \left[1 - \frac{1}{2} m_d (1 + \cos \Omega t) \right] \tag{4}$$

The angular frequency at resonance is $\omega_0 = \gamma B_0$, where B_0 is the external magnetic field magnitude; $\omega = 2\pi f$, with f and γ being the frequency of the microwave radiation and the free-electron gyromagnetic ratio, respectively. The Bloch equations in the rotating frame (Abragam & Bleaney, 1970) applicable to a spin packet at frequency $\Delta\omega$ ($= \omega_0 - \omega$) off resonance, under the action of an amplitude-modulated microwave magnetic field described by eq. (3) are:

$$\frac{\partial m}{\partial t} = -\left(i\Delta\omega + \frac{1}{T_2}\right)m + \frac{i\gamma B_1}{(1+m_c)}(1+m_c\cos\Omega t)M_z \tag{5}$$

$$\frac{\partial M_z}{\partial t} = -\frac{\gamma(m-m^*)}{2i}\frac{B_1}{(1+m_c)}(1+m_c\cos\Omega t) - \frac{M_z - M_0}{T_1} \tag{6}$$

In (5) and (6), $m = M_x + iM_y$, and M_0 is the equilibrium magnetization along the z-axis. It is noted that over and above eqs. (5) and (6), the other effects that influence the measurement of relaxation times (*e.g.*, unresolved hyperfine structure, overall spectral lineshape, spectral diffusion) are not taken into account here.

In normal field-modulated spectra the modulation is kept smaller than the line width, and in careful work the effect of the modulation on the line shape is considered. However, the sidebands at 100 kHz or less are almost always within the envelope of the line shape, except for very narrow lines. The fast modulation that is required to obtain a relaxation time-dependent response in modulation spectroscopy inherently has more widely spaced sidebands. For example, at 200 MHz modulation the sidebands are approximately 71 G away from the center band. The modulation index and the d.c. offset can be varied to obtain a wide range of modulation conditions (*cf.* Haworth & Richards, 1966; Losee *et al.*, 1997; Reference Data for Radio Engineers, 1968). It should be noted that the finite bandwidth caused by resonator Q can attenuate effects of sidebands and resultant signal amplitude as the modulation frequency is increased. This calculation is beyond the scope of this chapter but needs to be performed for a given experimental arrangement in order to be able to use the predictions of this chapter to actually measure relaxation times.

3.2. Case 1: Condition of Negligible Saturation

When saturation factor $a = \mu_0 H_1^2 \gamma^2 T_1 T_2 \ll 1$, the differential expression for the emf induced in a pickup coil along the z-axis (eq. (1)) is solved by applying Fourier expansions for m and M_z at resonance in Bloch's equations (5) and (6). This leads to the following expression for the signal:

$$S(X) = S_0 X \left[\frac{1 + p^2 X^2/4}{(1+X^2)(1+p^2 X^2)}\right]^{1/2} \tag{7}$$

where $X = \Omega T_1$, $p = T_2/T_1$, and $S_0 = (\mu_0 n Q_b m_c a M_0)/[RT_1(1+m_c^2)]$.

The normalized signal, S/S_0, as obtained from eq. (7) is plotted in Figure 3 for two values: $p = 1.0$ (denoted as "large") and $p \ll 1.0$ (negligible) as functions of X ($= \Omega T_1$). The data show that S/S_0 is biphasic with respect to X, with an asymptotic limit of approximately 0.5 that increases to 1.0 when $p \ll 1$ (note that p is always less than 1). The tangents for $X = 0$ and the asymptote intersect at $X = 0.5$ and 1.0, respectively, for the two cases. This provides a graphical solution for determining the value of T_1 as shown in Figure 3.

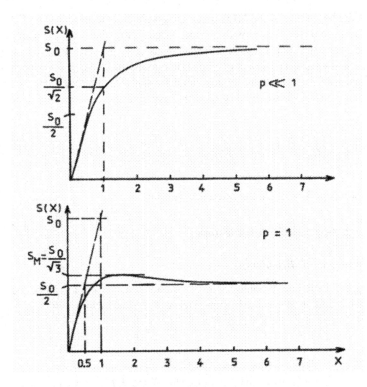

Figure 3. Graphical method of determining T_1 times for the case p ($= T_2/T_1$) $\ll$ 1 (negligible) and $p = 1.0$ (after Fretier, 1979; Vergnoux, 1996).

The spin–lattice relaxation times accessible to measurement by the modulation technique depend on one's choice of the modulation frequency, $\Omega/2\pi$. For example, a typical value of $\Omega/2\pi$ used for modulation and detection is 1 MHz. In the graphical determination of T_1 (Figure 3), the asymptote intersects the tangent to S/S_0 at small $X = \Omega T_1 = 0.5$ for values of p which are not too small, and this yields $T_1 = 10^{-6}/(4\pi) \approx 10^{-7}$ s. In order to plot S/S_0 vs. X, one needs to vary $\Omega/2\pi$ from $(0.33/4\pi) \times 10^{-6}$ to $(3.0/4\pi) \times 10^{-6}$, that is, in which case one should now have 30 kHz $< \Omega/2\pi < 17$ MHz. Under these conditions, the resultant T_1 ($=(2\Omega)^{-1}$) are in the range 10^{-6}–10^{-8} s.

The lower limit here can be extended downward by a factor of 100, that is, to 10^{-10} s, by using the fact that the reflection type EPR signal, S_{EPR}, as measured in a typical spectrometer does not vary with modulation frequency $\Omega/2\pi$, and therefore represents the asymptotic value experimentally. Furthermore, when $T_1 = 10^{-10}$ s it is not possible to obtain the asymptotic value because the requisite frequency becomes too high. At this point one is in the limit $X = \Omega T_1 \ll 1$ as T_1 is very small. The observable curve for $S(X)$ is then confined to the tangent part near the origin (small X, Figure 3). With the frequencies that are available in practice, one is then always at the beginning of the $S(\Omega)$ curve for graphical determination of T_1. To this end, one can use a free radical with a rather large T_1 together with the sample to be studied and measure their respective EPR signals (*i.e.*, $\overline{S}^R$ and $\overline{S}$) at a low modulation frequency (low X). Now it can be shown that ratio $\overline{S}/S_{EPR}$ is independent of the properties of the sample and is the same for both the reference sample and the sample being investigated. In this case the asymptotic value for the sample becomes

$$\overline{S} = \left(\frac{\overline{S}^R}{S_{EPR}^R}\right) S_{EPR}$$

where superscript R refers to the reference sample.

3.3. Case 2: Non-negligible saturation

3.3.1. Homogeneous Broadening

When the EPR line is homogeneously broadened, the signal can be expressed assuming negligible spectral diffusion (Hervé & Pescia, 1960, 1963a,b; Pescia & Hervé, 1963) as

$$S(X) = S_0 \frac{a}{1+a}\left(\frac{X(1+p^2X^2/4)^{1/2}}{\left[a^2+2a(1-pX^2)+(1+X^2)(1+p^2X^2)\right]^{1/2}}\right) \tag{8}$$

For this case, the tangent for $X \sim 0$ and the asymptote cross at point $X = (1 + a)/2$. T_1 is then found by extrapolating the curves drawn for different values of a to $a \rightarrow 0$, from which one deduces $T_1 = \Omega/2$. This is shown in Figure 4.

3.3.2. Inhomogeneous Broadening

When the EPR line is inhomogeneously broadened (Hervé & Pescia, 1960, 1963a,b; Pescia & Hervé, 1963; Ablart & Pescia, 1980; Ablart, 1978), there are many spin packets to be considered, each one of which is characterized by a Lorentzian lineshape that is much narrower than the composite line. The signal for a single spin packet can be expressed as a function of X:

$$S(X,\delta) = -\frac{m_c M_0 a}{T_1}\left|\left|\frac{1+ipX/2}{1+a+\delta^2 p^2 T_1^2}\times iX\left(1+\frac{\delta^2 p^2 T_1^2}{1+ipX}\right)\div\left\{a+(1+iX)\left(1+ipX+\frac{\delta^2 p^2 T_1^2}{1+ipX}\right)\right\}\right|\right| \tag{9}$$

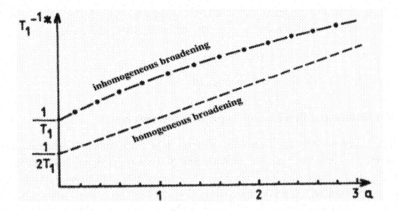

Figure 1. Determination of T_1 times by extrapolation for the case of non-negligible saturation for homogeneously and inhomogeneously broadened lines (after Fretier, 1979).

where $\delta = \omega - \omega_0$ is the difference between the resonance frequency of one packet from that of the whole line. Taking into account all the spin packets together distributed all over the composite line shape, one now has for the signal:

$$S(X,\omega) = \int_{-\infty}^{+\infty} S(X,\delta)f(\omega_0)d\omega_0 \tag{10}$$

In eq. (10), $f(\omega)$ is the shape function of a spin packet. When the line is purely inhomogeneous:

$$S(X,\omega) = f(\omega)\int_{-\infty}^{+\infty} S(X,\delta)d\omega_0$$

which becomes, using the method of residues,

$$S(X) = -\frac{\pi m_c M_0}{T_1 T_2} f(\omega)\big|S_0(X)\big|$$

with

$$S_0(X) = \frac{iaX(1+ipX/2)}{a(1+ipX)+(1+iX)(2ipX-a-p^2X^2)} \times$$

$$\left[\frac{-a+ipX}{(1+a)^{1/2}} + \frac{a+ipX-pX^2}{\{a(1+iX)/(1+ipX)+(1+iX)^2\}^{1/2}}\right] \tag{11}$$

When using the graphical method to estimate T_1, eq. (11) gives the condition where the asymptote intersects the tangent for small values of X at X_c to be

$$X_c = \Omega_c T_1 = (1+a)[1+(1+a)^{1/2}]/(2+a)$$

which yields $T_1 = 1/\Omega_c$ when extrapolated to $a \to 0$.

3.4. Sensitivity and Precision in Typical T_1 measurements

The benchmark for the technique is taken to be the number of centers required to observe an EPR line that is 1 G wide and possessing a signal-to-noise ratio of unity. Typical values obtained at 77 K using the various spectrometers operating with a pickup coil are: 5×10^{15} (0.2 GHz), 10^{14} (0.7 GHz), 5×10^{13} (4.0 GHz), and 10^{12} (8.2–12.4 GHz). In practice, one needs about 20 times these numbers to make a correct measurement of T_1.

The precision that can be obtained in these measurements is about 4% in the range 10^{-8} s $< T_1 < 10^{-6}$ s. It changes to 15% for very short T_1 times in the range 10^{-10} s $< T_1 < 10^{-8}$ s.

4. MATRIX TECHNIQUE TO SOLVE BLOCH'S EQUATIONS IN A ROTATING FRAME USING FOURIER-SERIES EXPANSION

In order to compare the signal induced in a pickup coil with the continuous wave signal detected in a resonator, one needs to find a common route towards the solution of the Bloch equations. The asymptotic solutions described in the preceding section are not applicable to the case in which the cw-EPR resonator is used. As an alternative, one may use a matrix technique.

The Fourier-series expansion of magnetic moments M_z and m in terms of magnetization moments $m_{(n)}$ and $M_{z(n)}$, and angular modulation frequency Ω, are:

$$m = \sum_{n=-\infty}^{\infty} m_{(n)} e^{in\Omega t} \tag{12}$$

$$M_z = \sum_{n=-\infty}^{\infty} M_{z(n)} e^{in\Omega t} \tag{13}$$

It is noted from (13) that $M_{(-n)}^* = M_{(n)}$, since M_z is real.

The moments of magnetization can be evaluated by the use of an appropriate matrix technique, which is described below. They can then be used to calculate the signals induced in a pickup coil, and the parallel and perpendicular components of the cw-EPR signals in a resonator.

In order to solve Bloch's equations using the matrix technique one introduces expansions (12) and (13) into Bloch equations (5) and (6), respectively. Using the relation $\cos \Omega t = \frac{1}{2}(e^{i\Omega t} + e^{-i\Omega t})$ and comparing the coefficients of $e^{in\Omega t}$ on the two sides of the equation leads to systems of coupled equations in terms of moments of expansion $m_{(n)}$ and $M_{z(n)}$. These are:

$$M_{z(n)} = \delta_{0n} M_0 + \frac{\gamma B_1}{2(1+m_c)(-n\Omega + i/T_1)} \times$$

$$\left[(m^*_{(-n)} - m_{(n)}) + \frac{m_c}{2} \left[\begin{matrix} (m^*_{-(n-1)} - m_{(n-1)}) + \\ (m^*_{-(n+1)} - m_{(n+1)}) \end{matrix} \right] \right] \tag{14}$$

In eq. (14), $\delta_{0n} = 1$ when $n=0$, and 0 otherwise;

$$M_{y(n)} = \frac{m^*_{-(n)} - m_{(n)}}{-2i} = \frac{\gamma B_1 \left[M_{z(n)} + \frac{m_c}{2} \{ M_{z(n-1)} + M_{z(n+1)} \} \right] F_n}{2i(1+m_c)} \tag{15}$$

where

$$F_n = \left\{ \frac{1}{n\Omega + \Delta\omega - i/T_2} + \frac{1}{n\Omega - \Delta\omega - i/T_2} \right\} \tag{16}$$

4.1. Signals Induced in a Pickup Coil vs. Resonator.

The signal that is induced in a pickup coil may be recovered by using a lock-in amplifier tuned to frequency Ω:

$$S_\alpha \text{ (pickup)} \propto \frac{\partial M_\alpha}{\partial t} = \Omega |M_{\alpha(1)}|; \quad (\alpha = z, y) \tag{17}$$

By contrast, the continuous wave signal that appears in a resonator is proportional to the steady-state (long-time) value of $M_{z,MAX} - M_{z,MIN}$:

$$S_\alpha \text{ (resonator)} \propto 2 |M_{\alpha(1)}|; \quad (\alpha = z, y) \tag{18}$$

One therefore needs the values of $M_{\alpha(1)}$ ($\alpha = x, y$) in order to calculate the signals. To accomplish this, the solutions of coupled equations (5) and (6) in $M_{\alpha(1)}$ ($\alpha = x, y$) are required. Substitution of (15) into (14) gives:

$$M_{z(n)} = \delta_{0n}M_0 + \frac{(\gamma B_1)^2}{2(1+m_c)^2(n\Omega - i/T_1)} \times$$

$$\left[\begin{array}{l} ZM_{z(n)}F_n + \dfrac{m_c}{2}\left(M_{z(n-1)}\{ZF_{n-1} + F_n\} + M_{z(n+1)}\{ZF_{n+1} + F_n\}\right) + \\[2mm] \left(\dfrac{m_c}{2}\right)^2\left(M_{z(n)}\{F_{n-1} + F_{n+1}\} + M_{z(n-2)}F_{n-1} + M_{z(n+2)}F_{n+1}\right) \end{array} \right] \tag{19}$$

whereas substitution of (14) into (15) yields

$$M_{y(n)} = \frac{(\gamma B_1)^2 F_n}{2(1+m_c)^2} \times$$

$$\left[\begin{array}{l} \dfrac{\left\{ZM_{y(n)} + \dfrac{m_c}{2}\left(M_{y(n-1)} + M_{y(n+1)}\right)\right\}}{(n\Omega - i/T_1)} + \dfrac{M_0\delta_{0n}(1+m_c)}{i(\gamma B_1)} \\[6mm] + \dfrac{m_c}{2}\left\{ \begin{array}{l} \dfrac{ZM_{y(n-1)} + \dfrac{m_c}{2}\left[M_{y(n-2)} + M_{y(n)}\right]}{(n-1)\Omega - i/T_1} + \dfrac{M_0\delta_{0(n-1)}(1+m_c)}{i(\gamma B_1)} \\[6mm] + \dfrac{ZM_{y(n+1)} + \dfrac{m_c}{2}\left[M_{y(n)} + M_{y(n+2)}\right]}{(n+1)\Omega - i/T_1} + \dfrac{M_0\delta_{0(n+1)}(1+m_c)}{i(\gamma B_1)} \end{array} \right\} \end{array} \right] \tag{20}$$

In (19) and (20), F_n is as defined by (16), and the expressions for $F_{n\pm1}$ can be obtained by substitution of $n \pm 1$ for n in the expression for F_n. It is noted that in (19) and (20) one substitutes $Z = 1$ when using the amplitude modulation expression given by eq. (3) (Herve & Pescia, 1960, 1963a,b; Pescia & Herve, 1963; Ablart & Pescia, 1980; Ablart, 1978). On the other hand, one substitutes $Z = 1 - \frac{1}{2}m_e$ and replaces (i) $(1+m_c)$ by 1 and (ii) $\frac{1}{2}m_c$ by $-\frac{1}{4}m_e$ when using the amplitude modulation expression given by eq. (4).

Equation (19) and (20) are the key equations that provide the recursion relations among $M_{\alpha(n+2)}$, $M_{\alpha(n+1)}$, $M_{\alpha(n)}$, $M_{\alpha(n-1)}$, $M_{\alpha(n-2)}$ $\{\alpha = z, y\}$ as expressed below by (21) and (22).

$$\begin{array}{l} A_{n,n+2}M_{z(n+2)} + A_{n,n+1}M_{z(n+1)} + \\[2mm] A_{n,n}M_{z(n)} + A_{n,n-1}M_{z(n-1)} + A_{n,n-2}M_{z(n-2)} = B_n \end{array} \tag{21}$$

$$\begin{array}{l} C_{n,n+2}M_{y(n+2)} + C_{n,n+1}M_{y(n+1)} + \\[2mm] C_{n,n}M_{y(n)} + C_{n,n-1}M_{y(n-1)} + C_{n,n-2}M_{y(n-2)} = D_n \end{array} \tag{22}$$

Equation (21) describes the recursion relation amongst $M_{z(n+2)}$, $M_{z(n+1)}$, $M_{z(n)}$, $M_{z(n-1)}$, $M_{z(n-2)}$, yielding a penta-diagonal matrix of complex elements. Similarly, (22) describes the recursion relation amongst $M_{y(n+2)}$, $M_{y(n+1)}$, $M_{y(n)}$, $M_{y(n-1)}$, $M_{y(n-2)}$.

One can express (21) as a matrix equation:

$$\begin{bmatrix} \text{matrix of} \\ \text{coefficients } A \end{bmatrix} \begin{bmatrix} \text{matrix of} \\ M_{z(n)} \end{bmatrix} = \begin{bmatrix} \text{matrix of} \\ B_n \end{bmatrix} \tag{23}$$

where, $B_n = \delta_{0n} M_0$. Likewise, from (22) one obtains the matrix equation

$$\begin{bmatrix} \text{matrix of} \\ \text{coefficients } C \end{bmatrix} \begin{bmatrix} \text{matrix of} \\ M_{y(n)} \end{bmatrix} = \begin{bmatrix} \text{matrix of} \\ D_n \end{bmatrix} \tag{24}$$

with

$$D_n = -i \frac{M_0 \gamma B_1}{2(1+m_c)} \left[\delta_{n0} + \left(\frac{m_c}{2} \right) \left\{ \delta_{0,n+1} + \delta_{0,n-1} \right\} \right]$$

The explicit expressions for the real and imaginary parts of coefficients A_{pq}, C_{pq}, B_n, and D_n are listed in Appendix 1.

The matrix equations given by (23) and (24) can be solved for $M_{z(n)}$ and $M_{y(n)}$, using the expressions derived in Appendix 2 to calculate the left-hand inverse of a complex matrix, which are then substituted back into (17) and (18) to calculate S_z (pickup), S_y (pickup), S_z (resonator), or S_y (resonator), respectively.

Although the matrices appearing in (23) and (24) are of infinite dimensions, they are, in practice, truncated for numerical calculations by checking for convergence so that the next term is less than a certain percentage of the sum up to the previous term. In the present calculations, terms containing $n = 0, \pm 1, \pm 2, \pm 3, \ldots,$ $\pm 11, \pm 12$ are retained (i.e., a 25×25 matrix) to permit calculation of signal intensity such that the contribution of the 26th term is less than 0.01%. It is seen that the solutions so obtained are consistent with those found numerically using the Runga-Kutta method (Press et al., 1992).

5. OUTLINE OF A LEAST-SQUARES FITTING PROCEDURE TO EVALUATE T_1 AND T_2 FROM PICKUP AND CW-EPR RESONATOR SIGNALS RECORDED AS FUNCTIONS OF Ω

The least-squares fitting (LSF) procedure that is used here is similar to that applied to the evaluation of spin-Hamiltonian parameters from cw-EPR line positions (cf. Misra, 1976, 1999). Relaxation times T_1 and T_2 are treated as components of parameter vector $\vec{p}$, to be estimated from the values of the measured signal, S_i^M, for various values of modulation frequency Ω_i ($i = 1, 2, \ldots, n$). In the present case, the χ^2 value required in the LSF procedure, a function of parameters T_1 and T_2, is expressed as

$$\chi^2 = \sum_{i=1}^{n} \left(S_i^C - S_i^M \right)^2 / \sigma_i^2 \tag{25}$$

In eq. (25), S_i^C, S_i^M, and σ_i are the calculated and measured values of the signal at modulation frequency Ω_i as given by (17) and (18), and the weighting factor (related to standard deviation) of the data point (measured signal), respectively. One can now obtain the values of the best-fit parameters in an iterative manner starting with judiciously chosen initial values of parameter vector $\vec{p}^I$ using the following equation:

$$\vec{p}^f = \vec{p}^I - \left\{ \left(D''^{-1} \right)_{p^I} \left(D' \right)_{p^I} \right\} \tag{26}$$

In eq. (26), D' is the column matrix, with the two components being $\partial \chi^2 / \partial p_j$ ($j = 1,2$), while D'' is the 2×2 matrix, with the four jk elements being $\partial^2 \chi^2 / \partial p_j \partial p_k$ ($j = 1,2$). The new set of parameters, $\vec{p}^f$, as obtained using eq. (26), are next used as initial values of parameters in place of $\vec{p}^I$ in (26), and the calculation is repeated, until such time that the new parameter values do not change significantly from the previous ones as revealed by the insignificant change in the resulting χ^2 value.

The first and second derivatives of χ^2 to be used in (26) are expressed as follows:

$$\frac{\partial \chi^2}{\partial p_j} = \sum_i 2 \frac{\left(S_i^C - S_i^M \right) \left(\partial S_i^C / \partial p_j \right)}{\sigma_i^2} \tag{27}$$

$$\frac{\partial^2 \chi^2}{\partial p_j \partial p_k} = \sum_i \left(\frac{2}{\sigma_i^2} \right) \left[\frac{\partial S_i^C}{\partial p_j} \frac{\partial S_i^C}{\partial p_k} + \left(S_i^C - S_i^M \right) \frac{\partial^2 S_i^C}{\partial p_j \partial p_k} \right] \tag{28}$$

In order to evaluate (27) and (28), the first and second derivatives of S_i^C with respect to the parameters are required. Since S_i^C is calculated using $M_{\alpha(n)}$; $\alpha = z, y$ as given by (19) and (20), the LSF calculation requires calculating the derivatives of $M_{0(n)}$ with respect to the parameters. The details are provided in Appendix 3.

6. ILLUSTRATIVE EXAMPLES

In order to illustrate the dependence of the modulation signals on spin–lattice (T_1) and spin–spin (T_2) relaxation times of spin packets, calculations are here carried out for a spin packet both on and off resonance to take into account inhomogeneous broadening. The values of the various quantities used for numerical calculations are as follows. At resonance, the external magnetic field value is $B_0 = 2\pi f/\gamma = 3283$ G at X-band ($f = 9.2$ GHz), using $\gamma = 1.7608 \times 10^7$ rad s^{-1}G^{-1}. The initial value of the magnetization is chosen to be that for 1 millimole of sample (number of $S =$

½ spins: $N=6\times10^{17}$ per cm^3) at room temperature ($T=295$ K). Under these conditions the Curie susceptibility, $\chi_0 = N\gamma^2\hbar^2 S(S+1)/(3kT) = 1.29\times10^{-9}$, resulting in $M_0=\chi_0 B_0=4.169\times10^{-6}$ G.

Three different sets of T_1, T_2, and B for the same coefficient of modulation, m_c (=0.1), and T_1 time (= 10^{-7} s) were used to illustrate the dependence of modulation signals on relaxation times at resonance, shown in Figures 5 and 6, respectively, for a pickup coil and resonator. Specifically, these values are:

Set a: $T_2=1\times10^{-7}$ s, $B_1=0.5$ G
Set b: $T_2=3\times10^{-8}$ s, $B_1=0.5$ G
Set c: $T_2=1\times10^{-9}$ s, $B_1=1.0$ G

6.1. Spin Packet at Resonance ($\Delta\omega = 0$)

It is seen from Figure 5, which illustrates the behavior of signal S_z (pickup) for each of these three sets of values, that indeed significantly different plots are obtained for different T_1, T_2 values. Furthermore, as the value of X (=ΩT_1) increases, each signal achieves an asymptotic constant value for large X. For the case when p (=T_2/T_1) << 1 (set **c**, $p = 0.01$), it is seen that the value of the signal increases monotonically with X, attaining an asymptotic value S_0 for large X that depends on values of T_1, T_2, B_1, m_c, which is consistent with the data illustrated in Figure 3. These data also show that the value of the signal is $S_0/\sqrt{2}$ when $X=1$.

When p is not too small (sets **a** and **b**), the signal first increases to achieve a maximum, after which its value decreases, finally attaining a constant value for large X. In this case: (i) the asymptotic value (large X) for S_z (pickup) is found to be $S_0/2$; (ii) the tangent of the S_z (pickup) versus X plot at very small values of X, where the curve is linear, intersects the extrapolated asymptotic value for large X at $X=0.5$, and (iii) the maximum value of the signal is $S_0/\sqrt{3}$. These observations are also consistent with those exhibited in Figure 3. As for signal S_y (resonator), it is seen from Figure 6 that for significant values of p, that is, $p=1.0$ (set **a**) and $p = 0.3$ (set **b**), the signal first increases with X up to about $X=1.213$ and $X=1.623$, respectively, and then drops off to a constant value with further increase in X, whereas for very small values of p, that is, for set (**c**) ($p = 0.01$), the signal decreases monotonically with increasing Ω, starting with the maximum value.

6.2. Spin Packet Off Resonance ($\Delta\omega \neq 0$)

Calculations of S_z (pickup) and S_y (resonator) were made for a spin packet for the same values of T_1, T_2, B_1, m_c, and M_0 as those for set (**a**) listed above, off-resonance by $\Delta\omega=20$ MHz. The resulting calculated signals are included in Figures 5 and 6, respectively. The required initial value of magnetization, M_0, off resonance is expected to be smaller than that at resonance, as it depends on deviation $\Delta\omega$ and the EPR lineshape (*e.g.* Gaussian, Lorentzian) depending on the properties of the sample. But in order to demonstrate the dependence of the signals on T_1, T_2 and facilitate easy comparison with the situation when the spin packet is at resonance,

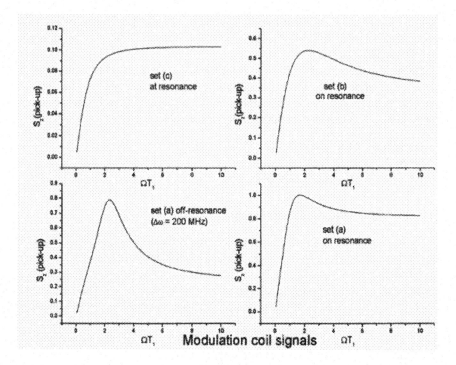

Figure 2. Plots of the signal induced in a pick-up coil, S_z, for a spin packet at resonance ($\Delta\omega = 0$) for the following set of values: (a) $T_1 = 10^{-7}$ s, $T_2 = 10^{-7}$ s, $B_1 = 0.5$ G; $m_c = 0.1$; (b) $T_1 = 10^{-7}$ s, $T_2 = 3 \times 10^{-8}$ s, $B_1 = 0.5$G; $m_c = 0.1$; (c) $T_1 = 10^{-7}$ s, $T_2 = 10^{-9}$ s, $B_1 = 1.0$G; $m_c = 0.1$; In addition, a plot is provided for a spin packet off resonance by 20 MHz calculated using the values of set (a).

the same value of M_0 was used in the calculation as that at resonance. This does not change any conclusions, since as seen from eqs. (17) and (18) the signals are directly proportional to M_0.

Figures 5 and 6 demonstrate that both S_z (pickup) and S_y (resonator) are quite sensitive to T_1, T_2. The relaxation times can thus be determined by using the LSF procedure to simultaneously fit a large number of data points for S_z (pickup) or S_y (resonator) obtained using various values of the modulation frequency. Furthermore, it is seen from Figures 5 and 6 that there already exist significant changes in the signals off resonance even by as small as ~0.3% from f ($\Delta\omega = 20$ MHz). Thus, in order to estimate T_1 and T_2 more accurately, additional data points, to be used in least-squares fitting, should be obtained by measuring signals of spin packets off resonance, achieved by setting the external magnetic field at an appropriate value.

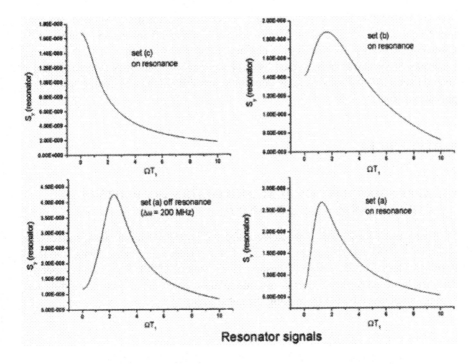

Figure 3. Plots of the signal induced in a pick-up coil, S_y, for a spin packet at resonance ($\Delta\omega = 0$) for the following set of values: (a) $T_1 = 10^{-7}$ s, $T_2 = 10^{-7}$ s, $B_1 = 0.5G$; $m_c = 0.1$; (b) $T_1 = 10^{-7}$ s, $T_2 = 3 \times 10^{-8}$ s, $B_1 = 0.5G$; $m_c = 0.1$; (c) $T_1 = 10^{-7}$ s, $T_2 = 10^{-9}$ s, $B_1 = 1.0G$; $m_c = 0.1$; In addition, a plot is provided for a spin packet off resonance by 20 MHz calculated using the values of set (a).

7. CONCLUDING REMARKS

The amplitude modulation technique has been used to measure spin–lattice relaxation times in the range 10^{-6}–10^{-10} s by using a pickup coil (Hervé & Pescia, 1960, 1963a,b; Pescia & Hervé, 1963; Ablart & Pescia, 1980; Ablart, 1978). This technique can be further improved with the help of modern technological advances to construct efficient microwave amplitude-modulated spectrometers, such as that using the crossed-field resonator. Alternatively, in the pickup coil technique one can use a dielectric resonator and place the pickup coil inside it (Forrer *et al.*, 2005). In this chapter, calculations are presented corresponding to the longitudinal (M_z) and transverse (M_y) signals that are induced in a pickup coil and, alternatively, the transverse signals (M_x and M_y) that might be detected by a resonator used in a cw-EPR experiment. Representative fast matrix calculations have been presented that demonstrate the dependence of the signal upon amplitude modulation using an arbitrary coefficient of modulation. Both T_1 and T_2 can be determined by simultaneously fitting the data corresponding to a set of experimentally applied modulation frequencies corresponding to spin packets on and off resonance. An accurate knowledge of relaxation times is necessary to unravel the nature of the relaxation processes in systems characterized by very short relaxation times of paramagnetic centres, such as amorphous silicon containing dangling bonds, spinels, glasses, and polymer resins.

8. APPENDICES

8.1. APPENDIX 1: LIST OF REAL AND IMAGINARY PARTS OF COEFFICIENTS A_{pq}, C_{pq}, B_r, AND D_r

The expressions below are derived for the amplitude modulation expression given by eq. (3), as defined by Hervé & Pescia (Hervé & Pescia, 1960, 1963a,b; Pescia & Hervé, 1963; Ablart & Pescia, 1980; Ablart, 1978). They can be specifically transformed to the case of strong modulation when the amplitude modulation expression is given by (4) by making the substitution

$$Z = 1 - \frac{m_d}{2}$$

replacing $(1 + m_c)$ by 1, and replacing $m_c / 2$ by $-m_d / 4$.

The real (Re) and imaginary (Im) parts of the matrix elements of matrix [A] in eq. (23) that are used in the modulation expression (eq. (3)) with $Z=1$:

$$G_n^{\pm} = \left(n \pm \frac{\Delta\omega}{\Omega}\right)\Omega \text{ , as:}$$

$$\text{Re}\left(A_{n,n\pm2}\right) = -\frac{\left(\gamma B_1\right)^2 \left(\dfrac{m_c}{2}\right)^2 T_1 T_2}{2\left(1+m_c\right)^2 Z\left(1+n^2\Omega^2 T_1^2\right)} \tag{8.1.1a}$$

$$\times \left[\frac{nG_{n\pm1}^{+}\Omega T_1 T_2 - 1}{1+G_{n\pm1}^{+2}T_2^2} + \frac{nG_{n\pm1}^{-}\Omega T_1 T_2 - 1}{1+G_{n\pm1}^{-2}T_2^2}\right]$$

$$\text{Im}\left(A_{n,n\pm2}\right) = -\frac{\left(\gamma B_1\right)^2 \left(\dfrac{m_c}{2}\right)^2 T_1 T_2}{2\left(1+m_c\right)^2 Z\left(1+n^2\Omega^2 T_1^2\right)} \tag{8.1.1b}$$

$$\times \left[\frac{n\Omega T_1 + G_{n\pm1}^{+}T_2}{1+G_{n\pm1}^{+2}T_2^2} + \frac{n\Omega T_1 + G_{n\pm1}^{-}T_2}{1+G_{n\pm1}^{-2}T_2^2}\right]$$

$$\mathrm{Re}\left(A_{n,n\pm1}\right) = -\frac{\left(\gamma B_1\right)^2 \left(\dfrac{m_c}{2}\right) T_1 T_2}{2(1+m_c)^2(1+n^2\Omega^2 T_1^2)}$$

$$\times \left[\frac{A'}{(1+G_n^{+2}T_2^2)(1+G_{n\pm1}^{+2}T_2^2)} + \frac{B'}{(1+G_n^{-2}T_2^2)(1+G_{n\pm1}^{-2}T_2^2)}\right] \qquad (8.1.2a)$$

In the preceding equations,

$$A' = \Omega T_1 T_2^3 n G_n^+ G_{n\pm1}^+ \left(ZG_n^+ + G_{n\pm1}^+\right) - T_2^2\left(ZG_n^{+2} + G_{n\pm1}^{+2}\right)$$
$$+\Omega T_1 T_2 n\left(G_n^+ + ZG_{n\pm1}^+ - (1+Z)\Omega\right)$$

$$B' = \Omega T_1 T_2^3 n G_n^- G_{n\pm1}^- \left(ZG_n^- + G_{n\pm1}^-\right) - T_2^2\left(G_n^{-2} + ZG_{n\pm1}^{-2}\right)$$
$$+\Omega T_1 T_2 n\left(G_n^- + ZG_{n\pm1}^- - (1+Z)\Omega\right)$$

$$\mathrm{Im}\left(A_{n,n\pm1}\right) = -\frac{\left(\gamma B_1\right)\left(\dfrac{m_c}{2}\right) T_1 T_2}{2\left(1+m_c\right)^2\left(1+n^2\Omega^2 T_1^2\right)}$$

$$\times \left[\frac{C'}{(1+G_n^{+2}T_2^2)(1+G_{n\pm1}^{+2}T_2^2)} + \frac{D'}{(1+G_n^{-2}T_2^2)(1+G_{n\pm1}^{-2}T_2^2)}\right] \qquad (8.1.2b)$$

with

$$C' = T_2^3 G_n^+ G_{n\pm1}^+ \left(ZG_n^+ + G_{n\pm1}^+\right) + \Omega T_1 T_2^2 n\left(ZG_n^{+2} + ZG_{n\pm1}^{+2}\right)$$
$$+T_2\left[G_n^+ + ZG_{n\pm1}^+ + (1+Z)n\Omega^2 T_1\right]$$

$$D' = T_2^3 G_n^- G_{n\pm1}^- \left(ZG_n^- + G_{n\pm1}^-\right) + \Omega T_1 T_2^2 n\left(ZG_n^{-2} + G_{n\pm1}^{-2}\right)$$
$$+T_2\left[G_n^- + ZG_{n\pm1}^- + (1+Z)n\Omega^2 T_1\right]$$

$$\mathrm{Re}\left(A_{n,n}\right) = 1 + \frac{\left(\gamma B_1\right)^2 T_1 T_2 E'}{2\left(1+m_c\right)^2\left(1+n^2\Omega^2 T_1^2\right)} \qquad (8.1.3a)$$

with

$$E' = \left\{ \frac{\left[1 - n\Omega G_n^{+} T_1 T_2\right]}{G_n^{+2} T_2^2 + 1} + \frac{\left[1 - n\Omega G_n^{-} T_1 T_2\right]}{G_n^{-2} T_2^2 + 1} \right\} Z$$

$$+ \left(\frac{m_c}{2}\right)^2 \left[\frac{1 - n\Omega G_{n+1}^{+} T_1 T_2}{G_{n+1}^{+2} T_2^2 + 1} + \frac{1 - n\Omega G_{n-1}^{+} T_1 T_2}{G_{n-1}^{+2} T_2^2 + 1} + \frac{1 - n\Omega G_{n+1}^{-} T_1 T_2}{G_{n+1}^{-2} T_2^2 + 1} + \frac{1 - n\Omega G_{n-1}^{-} T_1 T_2}{G_{n-1}^{-2} T_2^2 + 1} \right]$$

$$\mathrm{Im}\left(A_{n,n}\right) = -\frac{\left(\gamma B_1\right)^2 T_1 T_2 F'}{2\left(1 + m_c\right)^2 \left(1 + n^2 \Omega^2 T_1^2\right)}, \tag{8.1.3b}$$

with

$$F' = -\left\{ \frac{\left[n\Omega T_1 + G_n^{+} T_2\right]}{G_n^{+2} T_2^2 + 1} + \frac{\left[n\Omega T_1 + G_n^{-} T_2\right]}{G_n^{-2} T_2^2 + 1} \right\} Z$$

$$+ \left(\frac{m_c}{2}\right)^2 \left[\frac{n\Omega T_1 + G_{n+1}^{+} T_2}{G_{n+1}^{+2} T_2^2 + 1} + \frac{n\Omega T_1 + G_{n-1}^{+} T_2}{G_{n-1}^{+2} T_2^2 + 1} + \frac{n\Omega T_1 + G_{n+1}^{-} T_2}{G_{n+1}^{-2} T_2^2 + 1} + \frac{n\Omega T_1 + G_{n-1}^{-} T_2}{G_{n-1}^{-2} T_2^2 + 1} \right]$$

The elements of the right-hand side of eq. (23) are

$$\mathrm{Re}\left(B_n\right) = \delta_{on} M_0$$
$$\mathrm{Im}(B_n) = 0$$

As for the real and imaginary parts of the matrix elements of matrix $[C]$ in eq. (24), they are listed below:

$$\mathrm{Re}\left(C_{n,n\pm2}\right) = -\left[\frac{\left(\gamma B_1\right)^2 \left(\frac{m_c}{2}\right)^2 T_1 T_2 A''}{2\left(1 + m_c\right)^2} \right]$$

$$\times \left(\frac{1}{\left(G_n^{-2} T_2^2 + 1\right)\left(G_n^{+2} T_2^2 + 1\right)\left(\left(n \pm 1\right)^2 \Omega^2 T_1^2 + 1\right)} \right) \tag{8.1.4a}$$

with

$$A'' = \left(n \pm 1\right) T_1 T_2 \left[\left\{G_n^{-2} T_2^2 + 1\right\} G_n^{+} + \left\{G_n^{+2} T_2^2 + 1\right\} G_n^{-}\right]$$
$$- \left[\left\{G_n^{-2} T_2^2 + 1\right\} + \left\{G_n^{+2} T_2^2 + 1\right\}\right]$$

$$\text{Im}\left(C_{n,n\pm2}\right) = -\left|\frac{\left(\gamma B_1\right)^2 \left(\dfrac{m_c}{2}\right)^2 T_1 T_2 B''}{2\left(1+m_c\right)^2}\right|$$

$$\times\left(\frac{1}{[G_n^{-2}T_2^2+1][G_n^{+2}T_2^2+1]\left(n\pm1\right)^2 \Omega^2 T_1^2+1]}\right) \tag{8.1.4b}$$

with

$$B'' = [\{G_n^{-2}T_2^2+1\}G_n^{+} + \{G_n^{+2}T_2^2+1\}G_n^{-}]T_2$$
$$+[\{G_n^{-2}T_2^2+1\}+\{G_n^{+2}T_2^2+1\}]\left(n\pm1\right)\Omega T_1$$

$$\text{Re}\left(C_{n,n\pm1}\right) = -\left|\frac{\left(\gamma B_1\right)^2 \left(\dfrac{m_c}{2}\right)T_1 T_2 C''}{2\left(1+m_c\right)^2 \left(1+n^2\Omega^2 T_1^2\right)}\right| \times$$

$$\left(\frac{1}{[G_n^{-2}T_2^2+1][G_n^{+2}T_2^2+1][\left(n\pm1\right)^2 \Omega^2 T_1^2+1]}\right) \tag{8.1.5a}$$

with

$$C'' = \Omega T_1 T_2[\{G_n^{-2}T_2^2+1\}G_n^{+} + \{G_n^{+2}T_2^2+1\}G_n^{-}]$$
$$\times\left[n\{\left(n\pm1\right)^2 \Omega^2 T_1^2+1\}+\left(n\pm1\right)\{n^2\Omega^2 T_1^2+1\}\right]$$
$$-[\{G_n^{-2}T_2^2+1\}+\{G_n^{+2}T_2^2+1\}]$$
$$\times\left(n\pm1\right)^2 \Omega^2 T_1^2+1+(n^2\Omega^2 T_1^2+1)Z]$$

$$\text{Im}\left(C_{n,n\pm1}\right) = -\left|\frac{\left(\gamma B_1\right)^2 \left(\dfrac{m_c}{2}\right)T_1 T_2 D''}{2\left(1+m_c\right)^2 \left(1+n^2\Omega^2 T_1^2\right)}\right| \times$$

$$\left(\frac{1}{[G_n^{-2}T_2^2+1][G_n^{+2}T_2^2+1][\left(n\pm1\right)^2 \Omega^2 T_1^2+1]}\right) \tag{8.1.5b}$$

with

$$D'' = \Omega T_1 [\{G_n^{-2}T_2^2 + 1\} + \{G_n^{+2}T_2^2 + 1\}]$$
$$\times [n\{(n\pm 1)^2 \Omega^2 T_1^2 + 1\} + (n\pm 1)\{n^2\Omega^2 T_1^2 + 1\}]$$
$$+ [\{G_n^{-2}T_2^2 + 1\}G_n^+ + \{G_n^{+2}T_2^2 + 1\}G_n^-]$$
$$\times \Omega T_2 [(n\pm 1)^2 \Omega^2 T_1^2 + 1 + (n^2\Omega^2 T_1^2 + 1)Z]$$

$$\mathrm{Re}(C_{n,n}) = 1 - \frac{(\gamma B_1)^2 T_1 T_2 E''}{2(1 + m_c)[G_n^{-2}T_2^2 + 1][G_n^{-2}T_2^2 + 1]} \tag{8.1.6a}$$

with

$$E'' = \left(\frac{Z}{n^2\Omega^2 T_1^2 + 1}\right) \times \left(\frac{n\Omega T_1 T_2 [\{G_n^{-2}T_2^2 + 1\}G_n^+ + \{G_n^{-2}T_2^2 + 1\}G_n^-] -}{\{G_n^{-2}T_2^2 + 1\} + \{G_n^{+2}T_2^2 + 1\}}\right)$$

$$+ \left(\frac{m_c}{2}\right)^2 \left\{ \frac{\left(\frac{(n+1)\Omega T_1 T_2 [\{G_n^{-2}T_2^2 + 1\}G_n^- + \{G_n^{+2}T_2^2 + 1\}G_n^-] -}{[\{G_n^{-2}T_2^2 + 1\} + \{G_n^{-2}T_2^2 + 1\}]}\right)}{([n+1]^2 \Omega^2 T_1^2 + 1)} + \frac{\left(\frac{(n-1)\Omega T_1 T_2 [\{G_n^{-2}T_2^2 + 1\}G_n^- + \{G_n^{+2}T_2^2 + 1\}G_n^-] -}{[\{G_n^{-2}T_2^2 + 1\} + \{G_n^{+2}T_2^2 + 1\}]}\right)}{([n-1]^2 \Omega^2 T_1^2 + 1)} \right\}$$

$$\mathrm{Im}(C_{n,n}) = -\frac{(\gamma B_1)^2 T_1 T_2 F''}{2(1 + m_c)(G_n^{-2}T_2^2 + 1)(G_n^{+2}T_2^2 + 1)} \tag{8.1.6b}$$

with

$$F'' = \left(\frac{Z}{n^2\Omega^2 T_1^2 + 1}\right) \times$$

$$\left(\frac{T_2[\{G_n^{-2}T_2^2 + 1\}G_n^+ + \{G_n^{+2}T_2^2 + 1\}G_n^-] +}{n\Omega T_1 [\{G_n^{-2}T_2^2 + 1\} + \{G_n^{+2}T_2^2 + 1\}]}\right)$$

$$+ \left(\frac{m_c}{2}\right)^2 \left\{ \frac{\left(\frac{T_2[\{G_n^{-2}T_2^2 + 1\}G_n^+ + \{G_n^{+2}T_2^2 + 1\}G_n^-] +}{(n+1)\Omega T_1 [\{G_n^{-2}T_2^2 + 1\} + \{G_n^{+2}T_2^2 + 1\}]}\right)}{([n+1]^2 \Omega^2 T_1^2 + 1)} + \frac{\left(\frac{T_2[\{G_n^{-2}T_2^2 + 1\}G_n^+ + \{G_n^{+2}T_2^2 + 1\}G_n^-] +}{(n-1)\Omega T_1 [\{G_n^{-2}T_2^2 + 1\} + \{G_n^{+2}T_2^2 + 1\}]}\right)}{([n-1]^2 \Omega^2 T_1^2 + 1)} \right\}$$

The real and imaginary parts of the right-hand side of (24) are

$$\text{Re}(D_n) = 0$$

$$\text{Im}(D_n) = -\frac{M_0 \gamma B_1}{2(1+m_c)}\left[\delta_{n0} + \left(\frac{m_c}{2}\right)\{\delta_{0,n+1} + \delta_{0,n-1}\}\right]$$

8.2. APPENDIX 2: INVERSION THE COMPLEX MATRIX $A = (A' + A'')$

Expressions are derived in this appendix to invert the matrices given by eqs. (23) and (24), with complex matrix elements to calculate moments $M_{z(n)}$ and $M_{y(n)}$ for a system of coupled linear equations.

The complex elements of matrix A, whose elements are constituted by the co-efficients of $M_{\alpha(n)}$ ($\alpha=z,y$) in (19) and (20), can be separated into real and imaginary parts: $A = A' + iA''$, where A' and A'' are real matrices, with elements $A_{mn} = A'_{mn} + iA''_{mn}$. Let $A^{-1}{}_L$ be the left inverse of matrix A, so that $A^{-1}{}_L = I$, where I is the unit matrix. It can be expressed as $A^{-1}{}_L = F' + iF''$. Then

$$(F' + iF'') \times (A' + iA'') = I + iO \tag{8.2.1}$$

where I and O are the unit and null matrices, respectively. By comparing the real and imaginary parts, eq. (8.2.1) leads to

$$F''A' + F'A'' = O \tag{8.2.2}$$

and

$$F'A' - F''A'' = I \tag{8.2.3}$$

From (8.2.2),

$$F'' = -F'A''A'^{-1} \tag{8.2.4}$$

Substituting (8.2.4) into (8.2.3) for F'', and solving for F', one obtains for the real part of matrix A^{-1}

$$F' = (A' + A''A'^{-1}A'')^{-1} \tag{8.2.5}$$

Then using (8.2.5) for F' in (8.2.4), one obtains for F'' the imaginary part of A^{-1}:

$$F'' = -(A' + A''A'^{-1}A'')^{-1} A''A'^{-1} \tag{8.2.6}$$

Finally, using (8.2.5) and (8.2.4), the left inverse of complex matrix A is expressed as

$$A^{-1}{}_L = (A' + iA'')^{-1} = (A' + A''A'^{-1}A'')^{-1}(I - iA''A'^{-1}) \tag{8.2.7}$$

In (8.2.7) all matrices on the right-hand side are real. For inversion of a real matrix, one can use the LU-decomposition algorithm (Press *et al.*, 1992).

Finally, the vector of moments to various orders can be calculated from (19) and (20) as follows:

$$
\begin{bmatrix} \text{column vector} \\ \text{of } M_{\alpha(n)} \end{bmatrix} = \begin{bmatrix} \text{matrix} \\ \text{of } A_L^{-1} \end{bmatrix} \begin{bmatrix} \text{column vector} \\ \text{of } B_{\alpha(n)} \end{bmatrix}; \alpha = z, y \tag{8.2.8}
$$

8.3. APPENDIX 3: DERIVATIVES OF MAGNETIC MOMENTS

In order to calculate the derivatives of $M_{\alpha(n)}$; $\alpha = z,y$, required for least-squares fitting of data, one begins by taking the derivatives of (23) and (24), which are formally the same. For example, it is noted that eq. (24) is formally expressed in matrix form as

$$
[C][M] = [D] \tag{8.3.1}
$$

where $[C]$, $[M]$, and $[D]$ are the matrices constituted by elements C_{pq}, $M_{\alpha(n)}$, and D_n. [The same manipulations as described below can be applied to eq. (23).]

Now, by taking the first and second derivatives of (8.3.1) with respect to parameters p_j (T_1 and T_2), and noting that D does not depend on the parameters, one obtains, after simplification, for the first and second derivatives of moments $M_{\alpha(n)}$; $\alpha = z, y$:

$$
\left[\frac{\partial M}{\partial p_j} \right] = -[C]^{-1} \left[\frac{\partial C}{\partial p_j} \right] [M] \tag{8.3.2}
$$

and

$$
\left[\frac{\partial^2 M}{\partial p_j \partial p_k} \right] = -[C]^{-1} \left\{ \left[\frac{\partial^2 C}{\partial p_j \partial p_k} \right] [M] + \left[\frac{\partial C}{\partial p_j} \right] \left[\frac{\partial M}{\partial p_k} \right] + \left[\frac{\partial C}{\partial p_k} \right] \left[\frac{\partial M}{\partial p_j} \right] \right\} \tag{8.3.3}
$$

It is noted here that matrix $[C]^{-1}$ has already been calculated when solving for the matrix of moments, $[M]$, as described in Appendix 2.

Finally, it is seen from (8.3.2) and (8.3.3) that the task of calculating the first and second derivatives of $M_{\alpha(n)}$ reduces to calculating the first and second derivatives of the elements of matrix $[C]$ with respect to the parameters listed in Appendix 1. Details of how to calculate these derivatives are described below.

8.3.1. Calculation of First and Second Derivatives $\left[\partial C / \partial p_j \right], \left[\partial^2 C / \partial p_j \partial p_k \right]$

As seen from the expressions given in Appendix 1, each element, f, of coefficient matrix C can be expressed as a sum of two terms; each term is expressed as a ratio wherein both the numerator and denominators are functions of $\Delta\omega$. The second term is obtained from the first term by replacing $\Delta\omega$ by $-\Delta\omega$. Thus,

$$f = \frac{y_n(\Delta\omega)}{y_d(\Delta\omega)} + \frac{y_n(-\Delta\omega)}{y_d(-\Delta\omega)} \tag{8.3.4}$$

Then the first derivative of each term can be expressed formally, suppressing $\Delta\omega$ in the notation, as

$$\frac{\partial f}{\partial p_j} = -\frac{1}{y_d}\left[f\frac{\partial y_d}{\partial p_j} - \frac{\partial y_n}{\partial p_j}\right] \tag{8.3.5}$$

and

$$\frac{\partial^2 f}{\partial p_j \partial p_k} = -\frac{1}{y_d}\begin{bmatrix} -\dfrac{1}{y_d}\dfrac{\partial y_d}{\partial p_k}\left(f\dfrac{\partial y_d}{\partial p_j} - \dfrac{\partial y_n}{\partial p_j}\right) \\[2mm] +\dfrac{\partial f}{\partial p_k}\dfrac{\partial y_d}{\partial p_j} + f\dfrac{\partial^2 y_d}{\partial p_j \partial p_k} - \dfrac{\partial^2 y_n}{\partial p_j \partial p_k} \end{bmatrix} \tag{8.3.6}$$

An inspection of the elements of $[C]$, as listed in Appendix 1, reveals that y_d are either the product of two, three, or four, terms $y_d = y_{d1}y_{d2}$, or $y_d = y_{d1}y_{d2}y_{d3}$, or $y_d = y_{d1}y_{d2}y_{d3}y_{d4}$. Then the first and second derivatives of y_d with respect to the parameters in (8.3.5) and (8.3.6) can be calculated by the chain rule, and expressed in terms of

$$y_{dn}, \frac{\partial y_{dn}}{\partial p_j}, \frac{\partial^2 y_{dn}}{\partial p_j \partial p_k}.$$

Similar considerations apply to y_n. It is then straightforward, although time consuming, to derive specific expressions for these derivatives, which are not listed here due to their being too lengthy. (They can be obtained from the author.)

9. ACKNOWLEDGEMENTS

I am deeply indebted to the late Professor Jacques Pescia of Université Paul Sabatier, Toulouse, France, for introducing me to the technique of modulation spectroscopy, with whom I collaborated extensively since 1993, within the framework of Coopération France-Québec. (Sadly, he passed away in 2001.) In addition, I am grateful to Professors Sandra Eaton and Gareth Eaton of the University of Denver, USA, for informing me of the crossed-loop resonator technique and pointing out the deficiencies of the pickup coil arrangement. Some sections of this chapter are based closely on unpublished materials that they provided me.

10. REFERENCES

Ablart G. 1978. Étude de la relaxation spin-réseau de certains sels ioniques à partir des dépendence $T_1^{-1}(H_0)$ et $T_1^{-1}(T)$. PhD dissertation, Université Paul Sabatier, Toulouse, France.

Ablart G, Pescia J. 1980. Magnetic field dependence of spin–lattice relaxation in three iron group salts. *Phys Rev B* **22**:1150–1162.

Abragam A, Bleaney B. 1970. *Electron paramagnetic resonance of transition ions.* Oxford: Oxford University Press.

Alger RS. 1968. *Electron paramagnetic resonance: techniques and applications.* New York: Interscience Publishers (inversion-recovery technique).

Forrer J, Schuutz H, Tschaggelar R, Simovic B, Granwehr J, Schweiger A. 2005. Novel EPR Detection schemes and new detectors. EPR Instrumentation Workshop, Milwaukee, WI.

Hervé J, Pescia J. 1960. Résonance Paramagnétique — Mesure du temps de relaxation T_1 par modulation du champ radiofréquence H_1 et détection de variation d'aimantation selon le champ director. *Compt Rend* **251**:665–667.

Hervé J, Pescia J. 1963a. Résonance Magnétique Électronique — théorie phénoménologique de la mesure du temps de relaxation T_1, utilisant un champ radiofréquence modulé en amplitude: vérification experimentale sur le diphényl-picryl-hydrazyl. *Compt Rend* **255**:2926–2928.

Hervé J, Pescia J. 1963b. Influence de la température de carbonisation sur le temps de relaxation spin-réseau de charbon en presence d'oxygène. *Compt Rend* **256**:5076–5079.

Lopez R. 1993. Improvement in measurement of spin-lattice relaxation time T_1 in electron paramagnetic resonance: application to diluted copper calcium acetate and a Fe_2O_3-doped borate glass. PhD dissertation, Université Paul Sabatier, Toulouse, France [English translation by SK Misra]. In *Biological magnetic resonance*, vol. 25. Ed C Bender, L Berliner. New York: Springer, 2006.

Misra SK. 1976. Evaluation of spin-Hamiltonian parameters from EPR data by the method of least-squares fitting. *J Magn Reson* **23**:403–410.

Misra SK. 1998. Role of exchange interaction in effecting spin–lattice relaxation: interpretation of data on Cr^{3+} in $Cu_{2+x}Cr_{2x}Sn_{2-2x}$ spinel and dangling bonds in amorphous silicon. *Phys Rev B* **58**:14971–14977.

Misra SK. 1999. Angular variation of electron paramagnetic resonance spectrum: simulation of a polycrystalline spectrum. *J Magn Reson* **137**:83–92.

Misra SK. 2005. Microwave amplitude modulation technique to measure spin–lattice relaxation times: solution of Bloch's equations by a matrix technique. *Appl Magn Reson* **28**:55–67.

Pescia J. 1965. La mesure de temps de relaxation spin-réseau. *Ann Phys* **10**:389–406.

Pescia J, Hervé J. 1963. Mesure des temps de relaxation spin-réseau dans le cas d'un élargissement inhomogène de la raie. *Compt Rend* **256**:4621–4624.

Pescia J, Misra SK, Zaripov M. 1999a. Evidence for spin-fracton relaxation in the polymer resin P4VP doped with Co^{2+}, Nd^{3+}, Yb^{3+}. *Phys Rev Lett* **83**:1866–1869.

Pescia J, Misra SK, Zaripov M, and Servant Y. 1999b. Spin–lattice relaxation in the polymer resin P4VP doped with of Cu^{2+}, Cr^{3+}, Mn^{2+}, and Gd^{3+} transition ions possessing weak spin–orbit coupling. *Phys Rev B* **83**:1866–1869.

Piasecki W, Froncisz W, Hyde JS. 1996. Bimodal loop-gap resonator. *Rev Sci Instrum* **67**:1896–1904.

Poole CP. 1982. Electron spin resonance: a comprehensive treatise on experimental techniques, 2nd ed. New York: Wiley.

Press WH, Telkolsky SA, Vetterling WT, Flannery BP. 1992. *Numerical recipes in fortran.* New York: Cambridge UP.

Rinard GA, Quine RW, Eaton SS, Eaton GR, Froncisz W. 1994. Relative benefits of over-coupled resonators vs inherently low-Q resonators for pulsed magnetic resonance. *J Magn Reson* **A108**:71–81.

Rinard GA., Quine RW, Ghim BT, Eaton SS Eaton GR. 1996a. Easily tunable crossed loop (bimodal) EPR resonator. *J Magn Reson* **A122**:50–57.

Rinard GA, Quine RW, Ghim BT, Eaton SS Eaton GR. 1996b. Dispersion and Superheterodyne EPR using a bimodal EPR resonator. *J Magn Reson* **A122**:57–63.

Rinard GA, Quine RW, Eaton GR. 2000. An L-band crossed-loop (bimodal) resonator. *J Magn Reson* **144**:85–88.

Rinard GA, Quine RW, Eaton GR, Eaton SS. 2002. 250 MHz crossed-loop resonator for pulsed electron paramagnetic resonance. *Magn Reson Eng* **15**: 37–46.

Standley KJ, Vaughan RA. 1979. *Electron spin relaxation in solids.* New York: Plenum.

Vergnoux D, Zinsou PK, Zaripov M, Ablart G, Pescia J, Misra SK, Rakhmatullin R, Orlinskii S. 1996. Electron spin–lattice relaxation of Yb^{3+} and Gd^{3+} ions in glasses. *Appl Magn Reson* **11**:493–498.

Weidner RT, Whitmer CA. 1952. Recording of microwave paramagnetic resonance spectra. *Rev Sci Instrum* **23**:75–77.

Zinsou PK, Vergnoux D, Ablart G, Pescia J, Misra SK, Berger R. 1996. Temperature and concentration dependences of the spin-latice relaxation rate in four borate glasses doped with Fe_2O_3. *Appl Magn Reson* **11**:487–492.

CHAPTER 2

IMPROVEMENT IN THE MEASUREMENT OF SPIN–LATTICE RELAXATION TIME IN ELECTRON PARAMAGNETIC RESONANCE

Robert Lopez[a]. Translated by Sushil K. Misra[b]

[a]*Laboratoire de Magnétisme et d'Electronique Quantique, Université Paul Sabatier Toulouse III, 31077 Toulouse CEDEX, France,* [b]*Physics Department, Concordia University, 1455 de Maisonneuve Boulevard West, Montreal, Quebec H3G 1M8, Canada*

1. INTRODUCTION

The spin–lattice, or longitudinal, relaxation time T_1 plays an important role in magnetic resonance because it provides significant information about the coupling of a paramagnetic ion with its environment via its dependence on such factors as temperature, frequency (Scott & Jefferies, 1962; Kurtz & Stapleton, 1980), spin concentration (Gill, 1962), and magnetic field (Albart & Pescia, 1980; Nogatchewsky *et al.*, 1977). But the measurement of electronic spin–lattice relaxation times is problematic because the times span the range from the very short (10^{-15} s) to the very long (1 s; *cf.* Pescia, 1966). The one microsecond spin–lattice relaxation time demarcates "short" from "long" relaxation times, which traditionally have each required their own methods of measurement. For example, long relaxation times are measured by using cw-EPR spectrometers to record spectra at multiple power levels near and under the condition of saturation; the spin–spin and spin–lattice relaxation times are then calculated from lineshape parameters. But the so-called short relaxation times are not measurable on the time scale of common cw-EPR instrumental detection methods. Short spin–lattice relaxation times are therefore measured by resorting to different (*i.e.*, transient) magnetic resonance techniques such as pulsed saturation, spin echo (*cf.* Poole & Farach, 1971), and amplitude modulation (Hervé & Pescia, 1960a,b).

This chapter is a partial translation of the doctoral thesis of Robert Lopez entitled, *"Amélioration de la mesure du temps de relaxation spin-réseau T_1 en résonance paramagnétique électronique: Application a l'acetat de cuivre calcium dilué et un verre boraté dopé Fe₂O₃,"* Paul Sabatier University, Toulouse, France (1993) with permission.

Faced with a broad range of prospective spin–lattice relaxation times, the investigator needs two types of spectrometers, a situation that is further complicated if multi-frequency measurements are required. Furthermore, the phenomenological descriptions of measurements made by cw and transient spectrometers differ, as they correspond to separate solutions to Bloch's equations. This chapter describes refinements of both instrumental and theoretical/computational techniques that facilitate the measurement of spin–lattice relaxation times.

To begin, Bloch's phenomenological equations will be solved via Laplace transformation, which yields a natural and generalized expression for T_1. This computational solution, in turn, is put into practice via two instrumental improvements that enable one to measure T_1 over its entire range. In doing so, we have devised a simple technique for measuring long T_1 using a conventional EPR spectrometer equipped with a fast sweep and, in addition, constructed a device that increases the sensitivity and reliability of the modulation spectrometer for measurements of short T_1. Finally, this chapter concludes with practical measurements of spin relaxation times in two disparate spin systems using the modulation spectrometer. The first, a sample of calcium cadmium acetate hexahydrate doped with copper, is ideal for testing the temperature dependence of T_1. This sample material provides a large signal and, due to its weak exchange interaction, is expected to follow the theoretical dependence of T_1 as predicted by Bloembergen and Wang (1954). It therefore seemed possible for us to examine the behavior of T_1 at high temperature, that is, in the vicinity of the Debye temperature. The second system, a borate glass doped with Fe_2O_3, for which the study of the temperature dependence $T_1(T)$ does not appear to have been carried out before, was interesting due to coupling of the spin with its environment by a mechanism not well understood as of yet.

2. RELAXATION TIMES VIA GENERAL SOLUTION OF BLOCH'S EQUATIONS

The spectroscopic dynamics problem was examined mathematically for the case of the (two-level) magnetic resonance transition by Bloch, who described the temporal evolution of the magnetization in terms of a first-order differential equation analogous to $dn/dt = -k(n - n_0)$, where n represents a time-dependent function that, in this case, represents a spin-state population difference. (In a two-level system and in the form written, n would represent the population difference between the ground and excited states and the solution of the differential equation would correspond to the time course of the decay to the ground state.) The solution to this first-order differential equation is an exponential function in which a time constant is introduced and attributed to a characteristic relaxation time that is denoted by T_1. In other words, k is proportional to T_1^{-1}. This time constant T_1 is called the spin–lattice relaxation time, and is defined as the rate at which the electrons return to thermal equilibrium due to coupling with the lattice.

2.1. Bloch's Differential Equations of Motion

The magnetic resonance experiment is defined by the interaction between an external magnetic field, H_0, and the microscopic spin magnetic moments of a sample material, μ. The direction of the static field determines the coordinate frame (z-axis defined parallel to field H_0), and the experimental observables become the macroscopic magnetizations projected parallel and perpendicular to the z-axis. The quantum mechanics dictates that within the applied field the microscopic magnetic moments are spatially quantized with respect to the z-axis, and therefore macroscopic magnetization $\overline{M}$ is a vector whose magnitude and direction is determined by the sum of microscopic moments $\overline{M_i}$ of the material extended over the total volume of the sample. In the case of a two-level model (the so-called spin-$\frac{1}{2}$ model), the individual moments are aligned parallel or antiparallel to the z-axis, and the corresponding (population) numbers, n_+ and n_-, are determined by Boltzmann statistics. Since the z-component of magnetization M_z will be determined by difference $n = n_+ - n_-$, the preceding analogy using differential $\frac{dn}{dt} \propto n$ becomes apparent, that is,

$$\frac{dM_z}{dt} = -\frac{M_z - M_0}{T_1}$$

where $M_0 = \chi H_0$ and represents the equilibrium value of the macroscopic magnetization. This process implies that the spins lose energy to the lattice characterized by spin–lattice relaxation time T_1.

This first-order differential model suffices to describe the dynamical behavior of M_z. From a different perspective the applied field torques the microscopic magnetic moments, causing them to precess about the z-axis with an angular velocity defined as $\omega = (g\beta H_0)/\hbar$. The resultant equation of motion for the magnetization of a system of free spins in a static magnetic field can be expressed as

$$\frac{d\vec{M}}{dt} = \gamma\, \vec{M} \times \vec{H}$$

where coefficient $\gamma = \frac{g\beta}{\hbar}$ represents the gyromagnetic ratio.

Since there is no quantization perpendicular to z, the microscopic magnetization projected onto this plane is randomly oriented, and therefore macroscopic magnetization $M_\perp = 0$. Each individual microscopic moment, however, is in actuality subject to a local field that differs slightly from the specified laboratory field because of spatial inhomogeneities in H_0 and (relatively) weak fields of magneto-chemical origin within the sample. The individual spins therefore precess with different angular velocities (designated as set $\{\omega_i | \omega_i = \frac{g\beta H_{loc}}{\hbar}\}$) and phase. The field local to a given spin magnetic moment leads to its precession about it. This results in a spread of the resonance line when the ensemble of spins is taken into account. This spread attributes a linewidth to the resonance line.

The spectroscopic dynamics is induced by subjecting the sample to a second oscillatory field, H_1, with angular velocity ω. This field affects the phase angle of the resonant precessing spins, causing their moments to become (very nearly) coherent, and rendering $M_\perp$ nonzero. The dynamics of $M_\perp$ is often faster than that of magnetization M_z, because this process does not involve exchange of energy with the lattice, and is characterized by spin–spin relaxation time T_2:

$$\frac{dM_\perp}{dt} = \frac{M_\perp}{T_2}$$

Vector quantity $\overline{M}$ may now be resolved into its three components — M_x, M_y, and M_z — incorporating both time constants, T_1 and T_2. When the amplitude of the radio-frequency field, H_1, is much smaller than that of static field H_0, Bloch's differential equations may be written in the laboratory frame of reference (x,y,z):

$$\begin{cases} \dfrac{dM_x}{dt} = -|\gamma|(\vec{H} \times \vec{M})_x - \dfrac{M_x}{T_2} \\[2mm] \dfrac{dM_y}{dt} = -|\gamma|(\vec{H} \times \vec{M})_y - \dfrac{M_y}{T_2} \\[2mm] \dfrac{dM_z}{dt} = -|\gamma|(\vec{H} \times \vec{M})_z - \dfrac{M_z - M_0}{T_1} \end{cases}$$

Time constants T_1 and T_2 appearing here in the relaxation terms are called the longitudinal and transverse relaxation times.

2.2. The Steady-State Solution of Bloch's Equations

Bloch's phenomenological equations were solved for the case of slow passage through the resonance, by which it is meant that the rate of varying field H_0 is small compared to the time required to acquire a spectroscopic data point (H_0 is assumed to remain constant over this period). In the laboratory frame of reference (x, y, z), magnetic moments μ are subjected to time-dependent field $\vec{H} = \vec{i}H_1\cos(\omega t) + \vec{j}H_1\sin(\omega t) + \vec{k}H_0$, and so the evolution of vector product $(\vec{H} \times \vec{M})$ in Bloch's equations gives rise to first-order differential equations

$$\begin{cases} \dfrac{dM_x}{dt} = -|\gamma|[M_z H_1 \sin(\omega t) - H_0 M_y] - \dfrac{M_x}{T_2} \\[2mm] \dfrac{dM_y}{dt} = -|\gamma|[H_0 M_x - M_z H_1 \cos(\omega t)] - \dfrac{M_y}{T_2} \\[2mm] \dfrac{dM_z}{dt} = -|\gamma|[M_y H_1 \cos(\omega t) - M_x H_1 \sin(\omega t)] - \dfrac{M_z - M_0}{T_1} \end{cases}$$

The differential equations that are derived according to the axial symmetry of the laboratory coordinate frame may be simplified by switching to a coordinate frame that rotates (about the z-axis). The rotating frame of reference (x', y', z) is associated with field H_1 of frequency ν, and rotates about field H_0 with angular velocity $\omega = 2\pi\nu$. The equation of motion of the magnetization[4] of a system of free spins can be expressed, in the rotating frame of reference, as

$$\frac{d\vec{\tilde{M}}}{dt} = \gamma \vec{\tilde{M}} \times \left(\vec{H} - \frac{\vec{\omega}}{|\gamma|} \right)$$

$\vec{\omega}$ is the vector associated with the change of basis.

In the rotating frame, the magnetization therefore precesses about an effective field, $\vec{H}_e$, that is the vector sum of field $\vec{H}$ and fictitious field $-\vec{\omega}/|\gamma|$. The equations of motion of magnetization $\tilde{M}$ of a system of spins subjected to effective field $\vec{H}_e$ in the rotating frame can then be expressed as:

$$\begin{cases} \dfrac{d\tilde{M}_x}{dt} = -\dfrac{\tilde{M}_{\tilde{x}}}{T_2} + \Delta\omega \ \tilde{M}_{\tilde{y}} \\[2mm] \dfrac{d\tilde{M}_y}{dt} = -\Delta\omega \ \tilde{M}_{\tilde{x}} - \dfrac{\tilde{M}_y}{T_2} - \omega_1 M_z \\[2mm] \dfrac{dM_z}{dt} = \omega_1 \tilde{M}_{\tilde{y}} - \dfrac{M_z - M_0}{T_1} \end{cases}$$

with $\omega_1 = |\gamma| H_1$ and $\Delta\omega = 2\pi\nu - |\gamma| H_0 = \omega - \omega_0$.

Steady-state solutions of this system of equations are obtained when the magnetization no longer changes, that is, when the derivative of the magnetization with respect to time is zero:

$$\frac{d\tilde{M}_{\tilde{x}}}{dt} = \frac{d\tilde{M}_{\tilde{y}}}{dt} = \frac{d\tilde{M}_{\tilde{z}}}{dt} = 0$$

resulting in the following solutions:

$$\tilde{M}_{\tilde{x}} = \frac{\Delta\omega |\gamma| H_1 T_2^2}{1 + (T_2 \Delta\omega)^2 + |\gamma|^2 H_1^2 T_1 T_2} M_o$$

$$\tilde{M}_{\tilde{y}} = \frac{|\gamma| H_1 T_2}{1 + (T_2 \Delta\omega)^2 + |\gamma|^2 H_1^2 T_1 T_2} M_o$$

$$\tilde{M}_{\tilde{z}} = \frac{1 + (\Delta\omega T_2)^2}{1 + (T_2 \Delta\omega)^2 + |\gamma|^2 H_1^2 T_1 T_2} M_o$$

When there is no saturation ($|\gamma|^2 H_1^2 T_1 T_2 \ll 1$), the power absorbed by the spin system is

---

[4] To facilitate presentation of calculations, the notations of Abragam ($\tilde{M}_{\tilde{x}}, \tilde{M}_{\tilde{y}}, M_z$) (Abragam, 1961) will be used instead of the original notations of Bloch (u,v,M_z) (Bloch, 1946).

$$P = -\vec{M} \cdot \frac{d\vec{H}_e}{dt} = M_o T_2 |\gamma| H_1^2 \frac{\omega}{1 + (\Delta\omega T_2)^2}$$

The derivative of this power with respect to the field possesses a maximum when

$$\Delta H_{1/2} = \frac{\Delta\omega_{1/2}}{|\gamma|} = \frac{1}{|\gamma| T_2}$$

and this relation defines the half-width at half-maximum of the absorption line.

During the course of a cw-EPR experiment DC magnetic field H_0 is slowly swept across the resonance to record the spectrum. Ideally, one should ensure that passage across the resonance condition remains adiabatic, which means that the rate of field sweep is sufficiently slow so as to maintain a constant angle, θ, between magnetization $\vec{M}$ and effective field $\vec{H}_e$ (*cf.* Figure 1). If the sweep is not performed under adiabatic passage, the interaction energy will be perturbed.

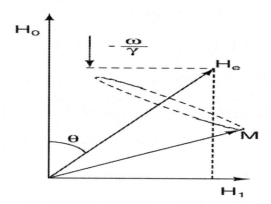

Figure 1. Precession of magnetization vector **M** about effective magnetic field H_0 in the rotating frame associated with field H_1.

Neglecting the relaxation terms, the equation of motion of the magnetization in the rotating frame associated with $\vec{H}_1$ can be expressed as:

$$\frac{d\vec{M}}{dt} = -|\gamma| \tilde{H}_e \times \vec{M}$$

According to the definition of adiabaticity given by Ehrenfest (Pake, 1962), adiabatic passage through the resonance will occur when

$$H_e^{-1} \left| \frac{dH_e}{dt} \right| << |\gamma| H_e$$

Since the effective field is equal to

$$\vec{H}_e = \left(H_o - \frac{\omega}{\gamma}\right)\hat{k} + H_1\vec{i}$$

the left-hand side of the preceding inequality is maximum when $H_e = H_1$ and $\omega = \omega_0$.

Thus, one can alternatively express the condition of adiabatic passage in the rotating frame as

$$H_1^{-1}\left|\frac{d\left(\vec{H}_o - \frac{\vec{\omega}}{k}\right)}{dt}\right| << |\gamma|H_1$$

Physically, the condition for adiabatic passage implies that effective magnetic field H_e should turn negligibly during one period of precession of the magnetization.

2.3. Experimental Methods of Measuring the Spin–Lattice Relaxation Time

From Bloch's equations and their solution there follows several experimental (*i.e.*, resonant) methods of measuring T_1. For example, it is a common practice to plot both the peak-to-peak amplitude and the peak-to-peak linewidth of the derivative EPR signal as a function of the square root of the microwave power $\sqrt{P}$ (*cf.* Poole, 1967). The square root of the microwave power is proportional to H_1, and so one has an experimental handle on the term $\gamma^2 H_1^2 T_1 T_2$ that appears in the denominator of the Bloch equations and is responsible for specifying the condition of saturation. The plot of signal amplitude *vs.* $\sqrt{P}$ is linear at low powers (non-saturating) and reaches a (derived) maximum value at $(1 + \gamma^2 H_1^2 T_1 T_2)^{-1} = 2/3$, from which it is possible to compute T_1 provided one knows both H_1 and T_2. The latter is obtained from the second plot of ΔH_{pp} *vs.* $\sqrt{P}$ that yields a low-power limiting (as $\sqrt{P} \to 0$) linewidth, ΔH_{pp}^0, that is proportional to T_2. This still leaves the problem of accurately determining H_1, which requires a difficult measurement of the radio-frequency field strength at the sample. This is typically done through the use of 2,2-diphenyl-1-picrylhydrazyl (DPPH) as a standard for comparative signal amplitude and linewidth *vs.* power studies (*cf.* Singer & Kommandeur, 1961).

The saturation method of determining T_1 has serious drawbacks, the most notable of which being the requisite measures of T_2 and H_1. Inhomogeneously broadened lines present further challenges to the measurement because how one defines the spin packet lineshape greatly affects the saturation behavior of the system (*cf.* Portis, 1953; Kittel & Abrahams, 1953). Lastly, the saturation factor, $(1 + \gamma^2 H_1^2 T_1 T_2)^{-1}$, will remain approximately unity (*i.e.*, non-saturating) when T_1 and T_2 are both very short, thereby rendering saturation H_1 inaccessible by cw-EPR spectrometers. And so time-domain EPR methods such as pulse saturation and electron spin echo are also used to measure T_1 via direct measures of the bulk magnetization following a short, saturating pulse of field H_1 (more facile than high-power cw-sources because of pulsed high power — *e.g.*, traveling wave tube — amplifiers that may be used with high spectral purity low-power sources).

The pulsed saturation method is modeled simply by our initial description of the state population difference,

$$\frac{dn}{dt} = \frac{n_0 - n}{T_1}$$

which has solution

$$n = n_0(1 - e^{-t/T_1})$$

In other words, a system of spins that have perturbed from their equilibrium state return to the equilibrium condition according to an exponential law characterized by time constant T_1. We shall use this fact shortly as a rationale for invoking the Laplace transform and associated transfer functions.

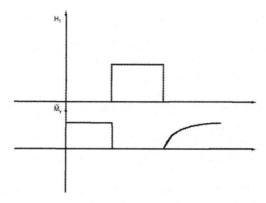

Figure 2. Temporal representation of the pulsed saturation experiment. The sample material is subjected to a saturating pulse, H_1, and the recovery of magnetization, M_y, with time constant T_1, is recorded.

A sample is set under the conditions of (non-saturating) resonance, and then subjected to a saturating pulse (Figure 2). During the saturating pulse, the spin populations equalize and transverse magnetization disappears. At the end of the pulse (*i.e.*, $\gamma^2 H_1^2 T_1 T_2 \ll 1$) the signal (transverse magnetization) recovers with time constant T_1.

Electron spin echo methods are similar to pulse saturation, but dispense with low-power monitoring field H_1, and instead use multiple pulses to refocus the dephasing spins so that a magnetization "echo" is detected at some time after the high-power pulse sequence. In the single-pulse saturation method described above, as time constant T_1 becomes short, it will become increasingly difficult to accurately record the discrete points (*i.e.*, EPR signal amplitude) along the exponential time scale; one needs a rapidly scanning boxcar averager or sampling oscilloscope with a very narrow gate window). But the spin echo methods allow one to temporally fix the response, which is in the form of an echo, so that its amplitude may be accurately measured by a boxcar signal averager. The temporal profile, in which the echo amplitude decays according to the same spin relaxation time constants, is recorded as a series of discrete steps as the interpulse spacing is increased.

It is apparent, however, that the three cited methods require different spectrometer configurations and have an optimal measurement range due to the technological limitations of the apparatus (response time of mixer diodes, rise/fall times of high power pulsed amplifiers, *etc.*). But, in principle, the magnetization recovery of the pulsed methods is attractive because it is conceptually simple and obviates the need for knowing H_1. This, in turn, motivates introducing a time-domain perturbation of H_1 and directly measuring the magnetization. Amplitude modulation methods of determining T_1 have been described (Halbach, 1954; Hervé & Pescia, 1960a,b; Look & Locker, 1968; Locker & Look, 1968) as an alternative to saturation and high-power pulse methods. Of these, the amplitude modulation method, when used with an alternative detection scheme (*i.e.*, a pickup coil to measure magnetization directly) provides the most versatile solution to measurement of T_1 over its entire range.

3. THE SOLUTION OF BLOCH'S EQUATIONS USING THE LAPLACE TRANSFORM

The method of determining T_1 via amplitude modulation of H_1 relies on variation of the modulation frequency, denoted by $\Omega/2\pi$, until it exceeds T_1^{-1}, at which point the magnetization cannot respond to the power variation and there is a loss in the detected EPR signal amplitude (Hervé & Pescia, 1960a). In a sense, this T_1 measurement is analogous to that used to analyze the impedance of a nonlinear system, such as a passive filter. The precept is that the response of a system $y(t)$ to some perturbation $x(t)$ is determined by some differential equation of order n. In the case of a linear system and perturbation $x(t)=A\,\sin(\omega t)$, one observes a response $y(t)=B\,\sin(\omega t+\phi)$ and one defines a transfer function as the ratio of output to input (in the frequency domain) $H(j\omega)=\left|H(\omega)\right|e^{j\phi}={}^{y(\omega)}\!/\!_{x(\omega)}$, where $y(\omega)$ and $x(\omega)$ are transforms of $y(t)$ and $x(t)$ in the frequency domain. Note that for this example describing a linear system $H(\omega)$ is independent of ω, and so a plot of $H(\omega)$ vs. ω is a straight line; resistance, for example, when $x(t)$ is a current and $y(t)$ is a voltage.

Transfer function $H(\omega)$, however, may be an explicit function of ω, as in the case of circuit impedance (again using the scenario described in the preceding paragraph). For example, when the system consists of a parallel RC circuit, the transfer function (again taken as the ratio of voltage out to current in) is given by the impedance, $\left|Z(\omega)\right|=({}^1\!/\!_R+jC\omega)^{-1}$, and so a plot of this $\left|H(\omega)\right|$ vs. ω consists of two linear portions of different slope, their intersection occurring at the resonance frequency of the system. The discontinuities in these plots of $\left|H(\omega)\right|$ vs. ω enable one to graphically determine system parameters (*i.e.*, time constants) and are known as Bode plots.

The preceding application of the transfer function is not limited to sinusoidal functions of $x(t)$ and $y(t)$, but are theoretically applicable to any type of perturbing signal by invoking the Fourier transform in order to generalize, $Y(\omega) = H(\omega)X(\omega)$, in which case the transfer function can be written

$$H(\omega) = \frac{Y(\omega)}{X(\omega)}$$

The same applies to functions transformed under the Laplace operational calculus:

$$H(p) = \frac{\mathcal{L}\{U_o(t)\}}{\mathcal{L}\{U_i(t)\}}$$

where $\mathcal{L}\{U_o(t)\}$ and $\mathcal{L}\{U_i(t)\}$ denote the Laplace transform (defined as $\mathcal{L}f(t) = F(p)$ $= \int f(t)\exp(-pt)dt$) of the output and input functions, respectively.

The Laplace transform and the transfer function provide the theoretical justification of Bode plots as a means to derive circuit parameters via the response of a circuit to some specified perturbation, for example, the analysis of passive filters. But the behavior of spin magnetization is analogous to that encountered when studying transient electrical phenomena. One can draw an analogy between the time-variant response of a spin system to that of an RC circuit: the descriptive differential equations are very nearly the same and the observable parameters are analogous (e.g., voltage is replaced by magnetization and the time constant RC by the spectroscopic time constant T_1). A Bode plot can therefore provide a graphical means of determining the time constants, namely T_1, from a spin system provided that the relevant transfer function obtained from the Bloch equations has the same parametric characteristics as in the electrical engineering analogue.

Consider, for example, the response of a spin system during a pulse-saturation experiment. A saturating pulse (H_1) at resonance is applied to the sample, after which H_1 is reduced (i.e., non-saturating, $\gamma^2 H_1^2 T_1 T_2 \ll 1$) and the recovery of magnetization recorded (§2.3, Figure 2). The variation of the magnetization of the sample due to a pulse of microwave field is described by the function

$$M(t) = M_{ns}\left(1 - e^{\frac{t}{T_1}}\right)$$

where M_{ns} is the equilibrium value of the non-saturated magnetization. The Laplace of $M(t)$ is

$$M(p) = \frac{T_1^{-1}}{\left(p + T_1^{-1}\right)}\frac{M_{ns}}{p}$$

This equation is identical to that which describes the charging of a capacitor through a resistor by applying a potential step denoted as V_e. In fact, the Laplace transform of voltage V_e across the terminals of the capacitor is equal to

$$V_c(p) = \frac{V_e}{p} \frac{\dfrac{1}{RC}}{p + \dfrac{1}{RC}}$$

This equation shows that the voltage across the capacitor increases exponentially toward the final value, V_e, with a time constant RC. In circuit theory, one describes the response to a step potential in terms of a transfer function defined by

$$H(p) = \frac{\dfrac{1}{RC}}{p + \dfrac{1}{RC}}$$

By replacing p by $j\omega$, the definition of V_c is changed from time domain (inverse Laplace transform) to frequency domain (Fourier transform). One then characterizes the electrical system by a representation in the Bode diagram of the modulus of the transfer function (cf. Figure 3):

$$|H(j\omega)| = \frac{\dfrac{1}{RC}}{\sqrt{\omega^2 + \left(\dfrac{1}{RC}\right)^2}}$$

The cutoff frequency of the transfer function is defined to be that corresponding to the point of intersection of the asymptote and the tangent to the curve describing the inverse time constant.

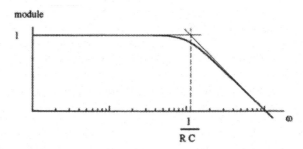

Figure 3. Bode diagram of $H(j\omega)$, the modulus of the transfer function of a series RC circuit.

3.1. Solution of Bloch's Equations in the Rotating Frame

The rotating frame $(\tilde{x}, \tilde{y}, z)$ associated with the radio frequency field is situated in such a way that the $\tilde{x}$ axis coincides with the axis of field H_1. As for the z-axis, it is common to both reference frames, and is coincident with static field H_0.

Bloch's equations can be expressed in matrix form as follows:

$$\frac{\partial}{\partial t}\begin{bmatrix} M_x \\ M_y \\ M_z \end{bmatrix} = -|\gamma|\begin{bmatrix} \tilde{M}_x \\ M_y \\ M_z \end{bmatrix}\begin{bmatrix} M_x \\ M_y \\ M_z \end{bmatrix}\times\begin{bmatrix} H_x \\ H_y \\ H_z \end{bmatrix} - \begin{bmatrix} \tau_2 & 0 & 0 \\ 0 & \tau_2 & 0 \\ 0 & 0 & \tau_1 \end{bmatrix}\begin{bmatrix} M_x \\ M_y \\ M_z \end{bmatrix} + \tau_1\begin{bmatrix} 0 \\ 0 \\ M_0 \end{bmatrix}$$

where parameters τ_1 and τ_2 are the inverses of the spin–lattice and spin–spin relaxation times, respectively.

In the moving frame, the differential equation of the first order can be expressed in the following form:

$$\frac{\partial\left[\vec{M}\right]}{\partial t} = \left[\vec{M}\right]\times\left(\left[\vec{\omega}\right]-|\gamma|\left[\vec{H}\right]\right)-\left[\tau_{1,2}\right]\left[\vec{M}\right]+\tau_1\left[\vec{M}_0\right]$$

Vector $\left[\vec{\omega}\right] = \begin{bmatrix} 0 \\ 0 \\ \omega_1 \end{bmatrix}$ is the vector representing the change of basis.

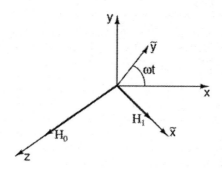

Figure 4. The rotating frame associated with the radio-frequency field.

In the moving frame (Figure 4), vector $\left[\vec{\tilde{H}}\right]$ is equal to $\begin{bmatrix} H_1 \\ 0 \\ H_0 \end{bmatrix}$, and vector

$\left[\vec{\tilde{M}}_0\right]$ is equal to $\begin{bmatrix} 0 \\ 0 \\ M_0 \end{bmatrix}$.

One can, therefore, factor the differential equation, which then reduces the time differential of the magnetization to

$$\frac{\partial \left[\vec{M}\right]}{\partial t} = [A]\left[\vec{M}\right] + \tau_1 \left[\vec{M}_0\right].$$

Matrix $[A]$ is a square matrix with elements

$$\begin{bmatrix} -\tau_2 & \Delta\omega & 0 \\ -\Delta\omega & -\tau_2 & -|\gamma|H_1 \\ 0 & |\gamma|H_1 & -\tau_1 \end{bmatrix}$$

and the definitions of the parameters are the same as those given in §2.2.

The Laplace transform of the system is equal to

$$\left(p[I]-[A]\right)\left[\vec{\tilde{M}}(p)\right] = \tau_1\left[\vec{\tilde{M}}_0(p)\right]$$

where $[I]$ is the unit matrix.

The solution of this equation is obtained by diagonalizing matrix $\left(p[I]-[A]\right)$ by using the Gaussian technique (cf. François et al., 1993) and square matrices $[Q_1]$ and $[Q_2]$ (Appendix I). The diagonalization of $\left(p[I]-[A]\right)$ by operation

$$[Q_2][Q_1]\left(p[I]-[A]\right)\left[\vec{\tilde{M}}(p)\right] = [Q_2][Q_1]\tau_1\left[\vec{\tilde{M}}_0(p)\right]$$

enables one to rewrite the system of equations in the form

$$\left[\vec{\tilde{M}}(p)\right] = [A']^{-1}[Q_2][Q_1]\frac{\tau_1\left[\vec{\tilde{M}}_0\right]}{p}.$$

Matrix $[A']=[Q_2][Q_1]\left(p[I]-[A]\right)$ is the diagonalized matrix.

The final form of the Laplace transform for the system of differential equations is

$$\begin{cases} \tilde{M}_{\tilde{x}} = -\dfrac{\Delta\omega\,\omega_h}{(p+\tau_1)\left[(p+\tau_2)^2+\Delta\omega^2\right]+\omega_h^2(p+\tau_2)}\dfrac{\tau_1 M_0}{p} \\[3ex] \tilde{M}_{\tilde{y}} = -\dfrac{\omega_h(p+\tau_2)}{(p+\tau_1)\left[(p+\tau_2)^2+\Delta\omega^2\right]+\omega_h^2(p+\tau_2)}\dfrac{\tau_1 M_0}{p} \\[3ex] M_z = -\dfrac{\Delta\omega^2+(p+\tau_2)^2}{(p+\tau_1)\left[(p+\tau_2)^2+\Delta\omega^2\right]+\omega_h^2(p+\tau_2)}\dfrac{\tau_1 M_0}{p} \end{cases}$$
,

with $\omega_h = |\gamma| H_I$.

The inverse transform of the solution is relatively complex. However, when $\omega_h^2 \ll \tau_1 \tau_2$ (unsaturated system), the common denominator becomes simple, being equal to

$$(p + \tau_1) \left[(p + \tau_2)^2 + \Delta\omega^2 \right]$$

One thus obtains the transient and steady-state solutions of Bloch's equations in the time domain as follows:

$$
\begin{cases}
\tilde{M}_{\tilde{x}}(t) = -\dfrac{\Delta\omega\, \omega_h}{\tau_2^2 + \Delta\omega^2} M_0 \\[2mm]
\qquad\quad - AM_0 \left[a_x e^{-\tau_1 t} + e^{-\tau_2 t} \left[a_C \cos(\Delta\omega t) + a_S \sin(\Delta\omega t) \right] \right] \\[4mm]
\tilde{M}_{\tilde{y}}(t) = -\dfrac{\omega_h \tau_2}{\tau_2^2 + \Delta\omega^2} M_0 \\[2mm]
\qquad\quad - AM_0 \left[a_y e^{-\tau_1 t} + e^{-\tau_2 t} \left[a_S \cos(\Delta\omega t) - a_C \sin(\Delta\omega t) \right] \right] \\[4mm]
M_z(t) = -M_0 \left(1 - e^{-\tau_1 t} \right)
\end{cases}
$$

with

$$A = \frac{\omega_h}{\left(\tau_2^2 + \Delta\omega^2 \right) \left[(\tau_1 - \tau_2)^2 + \Delta\omega^2 \right]}$$

and

$$
\begin{cases}
a_x = -\Delta\omega \left(\tau_2^2 + \Delta\omega^2 \right) \\
a_y = -(\tau_2 - \tau_1)\left(\tau_2^2 + \Delta\omega^2 \right) \\
a_C = \Delta\omega \tau_1 \left(2\tau_2 - \tau_1 \right) \\
a_S = \tau_1 \left(\tau_2^2 - \tau_1 \tau_2 - \Delta\omega^2 \right)
\end{cases}
$$

The linewidth of the EPR spectrum of a material provides the value of spin–spin relaxation time T_2 directly. On the other hand, the determination of spin–lattice relaxation time T_1 requires methods of indirect measurements.

3.2. The Second-Order Transfer Function for a System at Resonance

At resonance $\Delta\omega$ is equal to 0 (*i.e.*, $|\gamma|H_0 = 2\pi\nu$), and this value leads to the disappearance of the $\tilde{x}$ component of magnetization. At resonance, the system of differential equation is expressed as

$$
\left|
\begin{aligned}
\tilde{M}_{\tilde{x}} &= 0 \\
\tilde{M}_{\tilde{y}} &= -\frac{\omega_h}{(p+\tau_1)\,(p+\tau_2)+\omega_h^{\,2}}\,\frac{\tau_1\chi\bar{H}_0}{p} \\
M_z &= -\frac{(p+\tau_2)}{(p+\tau_1)\,(p+\tau_2)+\omega_h^{\,2}}\,\frac{\tau_1\chi\bar{H}_0}{p}
\end{aligned}
\right.
$$

When one applies the final-value theorem (the equations are multiplied by p, and then one takes the limit as p approaches 0) used in the theory of electric circuits, the usual values of magnetization are obtained:

$$
\left|
\begin{aligned}
\tilde{M}_{\tilde{x}} &= 0 \\
\tilde{M}_{\tilde{y}} &= -\frac{|\gamma|\,H_1 T_2}{1+|\gamma|^2\,H_1^{\,2}T_2 T_1}\,M_0 \\
M_z &= -\frac{1}{1+|\gamma|^2\,H_1^{\,2}T_2 T_1}\,M_0
\end{aligned}
\right.
$$

The change of $\tilde{M}_{\tilde{y}}$ with time appears in a form that is well known in the theory of electric circuits:

$$
\tilde{M}_{\tilde{y}} = -\frac{\tau_1\omega_h}{p^2+p(\tau_1+\tau_2)+(\tau_1\,\tau_2+\omega_h^{\,2})}\,\frac{M_0}{p}
$$

The first fraction in this equation represents a transfer function of the second order.

This solution to the Bloch equations using the Laplace transformation and the resultant form of M_y in the rotating frame is analogous to the problem of a double RC unit low-pass filter (*cf.* Figure 5). The voltage V_S corresponds to the output signal obtained from the filter in response to a voltage step V_e, that response being

$$
V_s(p) = \frac{\dfrac{1}{R^2 C^2}}{p^2+p\left(\dfrac{3}{RC}\right)+\left(\dfrac{1}{R^2 C^2}\right)}\,\frac{V_e}{p}
$$

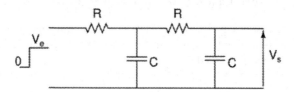

Figure 5. A double RC filter circuit, whose second-order transfer function representation models the response of a spin system in a pulsed saturation experiment.

It follows that, by analogy, a paramagnetic material placed in a static magnetic field and excited by a radio-frequency field of frequency

$$v = \frac{|\gamma| H_0}{2\pi}$$

exhibits a behavior similar to that of a low-pass filter of second order, with the difference here being that the variables are those of magnetization rather than those of voltage. The response of this filter to a step of static magnetic field H_0 is observed under these conditions.

3.3. Parameters of the Second-Order Transfer Function for Magnetic Resonance Calculations

The general formula of a second-order transfer function in the frequency domain is

$$H(j\omega) = \frac{K}{\omega_0^2 - \omega^2 + j2\zeta\omega_0\omega}$$

where ω_0 denotes the undamped characteristic frequency of the system, and ζ is the damping factor. We shall define a cutoff frequency, ω_C, as the frequency at which the absolute value of the transfer function is equal to that evaluated at $\omega=0$ divided by $\sqrt{2}$; this also corresponds to the familiar point at which the output signal is attenuated by $-3\,\mathrm{dB}$.

The transfer equation's parameters, as defined in a magnetic resonance experiment are identified as

$$K = \frac{\tau_1 \omega_h}{\omega_0^2} : \quad \text{gain at zero frequency}$$

$$\omega_0 = \sqrt{\tau_1 \tau_2 + \omega_h^2} : \quad \text{characteristic angular frequency}$$

$$\zeta = \frac{1}{2} \frac{\tau_1 + \tau_2}{\sqrt{\tau_1 \tau_2 + \omega_h^2}} : \quad \text{damping factor}$$

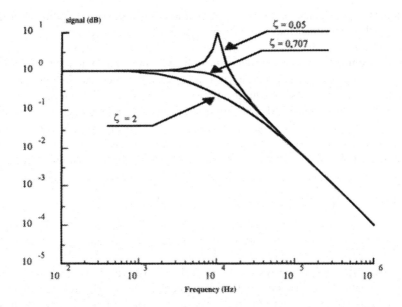

Figure 6. Variation of the transfer function as a function of the damping coefficient (Bode diagram).

The frequency-dependent behavior of the transfer function (*i.e.*, the Bode diagram) takes on a characteristic shape that is largely determined by the damping factor, as illustrated in Figure 6. When $\zeta = 2^{-1/2}$ (=0.707), the second-order transfer function resembles a first-order transfer function, and the Bode diagram consists of two linear portions of slope 0 and −40 dB/decade separated by an inflection point. For higher values ζ, the overall shape of the Bode diagram does not change, but the inflection point previously seen expands and assumes a characteristic slope of −20 dB per decade.

When the damping coefficient is less than $2^{-1/2}$, the shape of the Bode diagram changes and there appears a peak at resonant frequency

$$\omega_R = \omega_0 \sqrt{1 - 2\zeta^2}$$

The corresponding Q factor, defined by the ratio of the amplitude of the absolute value of the transfer function at angular frequency $\omega = \omega_R$ and that at angular frequency $\omega = 0$, is equal to

$$\frac{1}{2\zeta\sqrt{1 - \zeta^2}}$$

In the time domain, the damping factor determines the manner in which the system will attain its final value in the time domain. The term "final" is used to indicate the response of a system to a perturbation at the end of a time interval that is much longer than that of the transients. When this coefficient is less than one, called critical value, the transients possess a periodic character: the shape of the response function becomes a damped sinusoidal function. When the damping coefficient is equal to or greater than the critical value, the behavior becomes aperiodic, and the system attains the equilibrium value in an exponential manner.

The implications of this behavior with respect to a paramagnetic material will now be considered. The spin–spin relaxation time is always less than or equal to the spin–lattice relaxation time, and so one always has $\tau_2 \geq \tau_1$. In the absence of saturation, for which $\sqrt{\tau_1 \tau_2} >> \omega_h$, the damping factor falls within the range

$$1 \leq \zeta \leq \frac{1}{2}\sqrt{\frac{\tau_2}{\tau_1}}$$

where the lower and upper limits of the inequality correspond to $\tau_2 = \tau_1$ and $\tau_2 >> \tau_1$, respectively.

Coefficient ζ is always greater than the critical value, which implies that: (1) the frequency response does not exhibit a peak in the transfer function, and (2) the transients are of the aperiodic type. These remarks are in conformity with the results obtained for pulsed saturation in which the magnetization grows exponentially towards its final value.

3.4. Characteristics of the Transfer Function

At the start of a pulse saturation experiment the sample is saturated, which is equivalent, in terms of parameter ω_k, to setting the value of the transfer function equal to zero, that is, implying zero resultant magnetization. This is analogous to establishing a short circuit between the terminals of the second capacitor in the double RC-filter model. After the saturating pulse the amplitude of the microwave field, H_1, does not saturate the material, and one then observes an exponential increase in the value of the magnetization.

In fact, and again drawing in analogy to the filter model, one observes a response by the spin system to a step of magnetization. Since the time constant is equal to T_1, a cutoff frequency equal to T_1^{-1} is found in the Bode plane.

In the absence of saturation ($\omega_h^2 << \tau_1 \tau_2$), the value of the characteristic frequency of the transfer function depends only on τ_1 and τ_2: $\omega_0 \approx \sqrt{\tau_1 \tau_2}$.

The corresponding gain, K, is equal to ω_h / τ_2, which depends only on T_2, the spin–spin relaxation time.

To complete the calculation, cutoff frequency ω_c of the transfer function remains to be determined at –3 dB. It is obtained by imposing condition

$$\left| F(j\omega_c) \right| = \frac{|F(0)|}{\sqrt{2}} = \frac{K}{\sqrt{2}}$$

The solution of this equation (Appendix I) provides the relation between ω_c and the various parameters of the system:

$$\omega_c^2 = -\frac{1}{2}\left[(\tau_1+\tau_2)^2 - 2\omega_0^2\right] + \frac{1}{2}\sqrt{\left[(\tau_1+\tau_2)^2 - 2\omega_0^2\right]^2 + 4\omega_0^4}$$

Taking into account the relative values of τ_1 and τ_2, one obtains

$\tau_1=\tau_2=\tau$: the cutoff frequency is given by $\omega_c = \tau\sqrt{-1+\sqrt{2}} \approx 0.64\tau_1$.

$\tau_1 \ll \tau_2$: the cutoff frequency is given by $\omega_c = \tau_1$.

The second value clearly indicates that the spin–lattice relaxation time, ordinarily measured in the time domain, appears in the frequency-domain data. The variation of the absolute value of the transfer function is summarized in the Bode diagram (Figure 7), plotted for different values of ratio τ_1/τ_2, keeping the value of τ_2 constant.

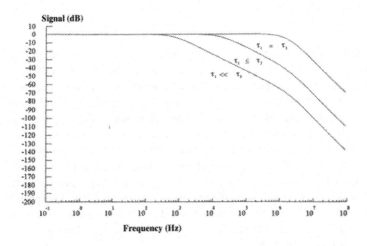

Figure 7. Variation of the cutoff frequency as a function of spin–lattice relaxation time T_1 (constant spin–spin relaxation time held constant).

3.5. Application of the Transfer Function to the Measurement of Spin–Lattice Relaxation Time T_1

The formalism of transfer functions will now be applied to the case of magnetic resonance. This theory shows that, in general, the system consisting of the sample at resonance is stable because all the poles of the transfer function possess real negative parts ($-\tau_1$ and $-\tau_2$), which is a necessary and sufficient condition.

Thus, when this system is perturbed by a step of magnetization, not only does the transfer function not diverge, it tends to a finite value.

We begin by representing the spin system by a device subjected to a sinusoidal signal of variable frequency (Figure 8). The Bode diagram is then given as the plot of output signal V_s as a function of frequency ω.

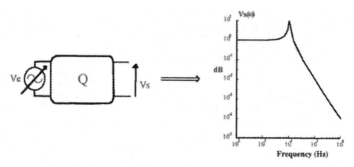

Figure 8. An experimental representation of how one determines the transfer function of a four-terminal network.

In the case of a conventional cw-EPR experiment, the field modulation applied along the z-axis may be varied, and one may measure the response of the system as the magnetization along the $\tilde{y}$ axis. But before we draw a Bode diagram for this type of experiment, we must first account for the resonance condition imposed by static field H_0 and specify those conditions under which the transfer function is described, that is, an experimentally valid description of the effective field.

The cw-EPR experiment is described in terms of a small-amplitude alternating field, $\tilde{h}_0$, that is superimposed upon static field H_0. The transfer function should then be plotted in the Bode plane. Figure 9 illustrates the change of effective field H_e as a function of amplitude modulation of the static field. The projection of dH_e/dt upon the $\tilde{y}$ axis is zero, whereas it does contribute to the variation of the z component of the magnetization. For this reason, only the behavior of magnetization $M_{\tilde{y}}$ has here been studied.

When the total field is described as $\bar{H}_0 + \tilde{h}_0 = \bar{H}_0 + \bar{h}_0 \cos(\Omega t)$ the denominator that appears in the transfer function, that is, $(p+\tau_1)\left[(p+\tau_2)^2 + \Delta\omega^2\right] + \omega_h^2(p+\tau_2)$ becomes

$$(p+\tau_1)\left[(p+\tau_2)^2 + \left(|\gamma|\bar{h}_0\right)^2\right] + \omega_h^2(p+\tau_2)$$

The amplitude of the field modulation must be sufficiently small so as to keep the system at resonance, and therefore

$$|\gamma|\tilde{h}_0 << \tau_1$$

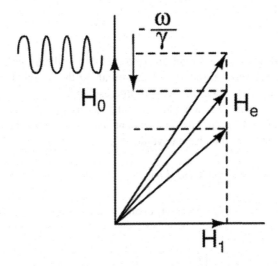

Figure 9. Variation of effective field H_e during the course of an amplitude modulation experiment.

Knowing that the half-width at half-maximum of the EPR absorption line is equal to $(|\gamma| T_2)^{-1}$, it is deduced that the amplitude of the alternating field must be much smaller than the width of the line.

Finally, the component along the $\tilde{x}$ axis, which is now nonzero, must be considered. Taking into account the preceding condition along with the condition of resonance in weak fields, transfer functions $F_{\tilde{x}}$ and $F_{\tilde{y}}$ of components $\tilde{M}_{\tilde{x}}$ and $\tilde{M}_{\tilde{y}}$ can be expressed as

$$
\begin{cases}
F_{\tilde{x}}(j\omega) = \dfrac{-|\gamma|\,\overline{h}_0\,\tau_1\,\omega_h}{(\tau_2 + j\omega)\left[\tau_1\,\tau_2 - \omega^2 + j\omega(\tau_1 + \tau_2)\right]} \\[4mm]
F_{\tilde{y}}(j\omega) = \dfrac{\tau_1\,\omega_h}{\tau_1\,\tau_2 - \omega^2 + j\omega(\tau_1 + \tau_2)}
\end{cases}
$$

The absolute values of these two functions for $\omega=0$ are equal to

$$
\begin{cases}
\left|F_{\tilde{x}}(0)\right| = \dfrac{|\gamma|\,\overline{h}_0\,\omega_h}{\tau_2^{\,2}} \\[4mm]
\left|F_{\tilde{y}}(0)\right| = \dfrac{\omega_h}{\tau_2}
\end{cases}
$$

One can now calculate the ratio of the amplitudes:

$$\frac{\left|F_{\tilde{y}}(0)\right|}{\left|F_{\tilde{x}}(0)\right|} = \frac{\tau_2}{\left|\gamma\right|\bar{h}_0}$$

Now, according to our assumption, $\left|\gamma\right|\bar{h}_0 \ll \tau_2$, which implies that $\left|F_{\tilde{y}}(0)\right| \gg \left|F_{\tilde{x}}(0)\right|$. The variation of the magnetization along the $\tilde{x}$ axis is negligible and will have no effect on the measurements of interest here.

Since the term in $\bar{h}_0$ can be neglected in the expression $\Delta\omega = \omega_0 - \left|\gamma\right|H_0$, matrix [A], defined in §3.1, remains independent of time. The Laplace transform of the differential equations then has the value

$$p\left[\vec{\tilde{M}}(p)\right] = [A]\left[\vec{\tilde{M}}(p)\right] + \tau_1\left[\vec{\tilde{M}}_0(p)\right]$$

It follows therefore that only vector $\mathbf{M}_0$ changes. The component along the z-axis is equal to

$$M_0(p) = \left|\gamma\right|H_0(p) = \left|\gamma\right|\left(\frac{\bar{H}_0}{p} + \bar{h}_0\frac{p}{p^2 + \Omega^2}\right)$$

With the help of matrices of transformation [Q_1] and [Q_2], the same as those given in §3.1, the response of the system of spins to the above perturbation is found to be

$$\tilde{M}_{\tilde{y}}(p) = \frac{\bar{M}_0}{p}F(p) + \bar{m}_0\frac{p}{p^2 + \Omega^2}F(p).$$

Here, parameter $F(p)$ and $\bar{m}_0$ are defined as

$$\begin{cases} F(p) = \dfrac{\tau_1\,\omega_h}{p^2 + p(\tau_2 + \tau_1) + (\tau_1\,\tau_2 + \omega_h^2)} \\ \bar{m}_0 = \left|\gamma\right|\bar{h}_0 \end{cases}$$

It is now possible to obtain the characteristic curve of the transfer function as long as the alternating field is smaller than the width of the absorption line at half-maximum, which is usually the case in an EPR experiment: the small value of the spin–spin relaxation time, T_2, causes the width of the lines to become greater than about 10 Gauss.

There are, however, practical limitations to this method of determining T_1. First, it must be remembered that the modulation coil and its associated parasitic capacitance behave as a resonant circuit. The frequency range over which the coil may be used will therefore be limited. Second, the receiver diode and the associated selective amplifier also limit the applicable frequency range. And, finally, the typical cavity quality factor of approximately 10^4 imposes a rather narrow bandpass to the measurement, typically 1 MHz. In practice, then, only spin–lattice relaxation times greater than 3 µs can be measured by the field modulation technique. The sensitivity of the method depends on the width at half-maximum, $\Delta H_{1/2}$, of the absorption line because the amplitude of the modulation field must be, at the most, $0.01\Delta H_{1/2}$.

4. AMPLITUDE MODULATION OF THE INCIDENT MICROWAVE FIELD AS AN ALTERNATIVE MEANS OF DETERMINING T_1: MEASUREMENT OF ULTRAFAST RELAXATION TIMES

4.1. Description of the Experiment

Analysis of the spin relaxation parameters by using the Bode diagram run into practical limitations when one uses the field modulation frequency as system perturbation. This limitation comes in the form of the frequency range over which the system can be perturbed, with the result being that one is constrained with respect to those relaxation times that may be measured. We therefore seek an alternative modulation scheme that is not as severely limited with respect to the applied frequency and so will be more generally applicable to the wide range of spin relaxation times. In this section we shall examine the method of H_1 (i.e., the resonant microwave field) amplitude modulation method and its use with pickup coils to measure the variation of M_z.

In a conventional EPR spectrometer, the H_1 field is perpendicular to DC field H_0, resulting in effective field H_e (Figure 10). If the H_1 field is now amplitude modulated at some frequency Ω, angle θ between static field H_0 and effective field H_e will temporally vary, which will likewise cause a temporal variation in the magnetization. The effect of amplitude modulating H_1 on H_e is confined to the so-called ρ-axis of Figure 10, but this effect on H_e will be manifest on M_z, making it possible to observe this effect on M_z by using a pickup coil whose cylindrical axis is oriented along H_0 because the projection of H_e along H_0 is unaltered. Returning now to the Bode diagram analysis, the electromotive force (emf) induced in the pickup coil will be measured in response to the frequency variation of an amplitude modulation applied to the incident microwave field under the condition of electron paramagnetic resonance.

We define the signal induced in the pickup coil as $S(\Omega)$, which will be proportional to dM_z/dt, that is

$$S(\Omega) = \frac{nQ_b\mu_0}{2R}\frac{dM_z}{dt}$$

The terms in the proportionality constant correspond to familiar coil parameters: n is the number of turns; Q_b is the quality factor; R is the radius; and μ_0 is the magnetic permeability of a vacuum.

The amplitude-modulated radio-frequency field is given by the equation

$$H_1(t) = H_1[1 + m\cos(\Omega t)]\cos(\omega t)$$

where m is the coefficient of modulation.

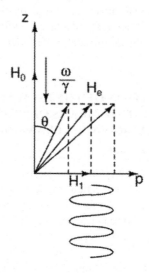

Figure 10. Vectorial representation of H_1 amplitude modulation.

The effect of amplitude modulation on magnetization is determined by substituting the modified expression for H_1 into Bloch's equations for the rotating field:

$$
\begin{cases}
\dfrac{d\tilde{M}}{dt} = i[1 + m\cos(\Omega t)]\,|\gamma|\,H_1\,\dfrac{M_z}{1+m} - \tau_2\tilde{M} \\[2ex]
\dfrac{dM_z}{dt} = -i[1 + m\cos(\Omega t)]\,|\gamma|\,H_1\,\dfrac{\tilde{M}^* - \tilde{M}}{2(1+m)} - \tau_1(M_z - M_0)
\end{cases}
$$

Magnetization M_z is expressed as a Fourier series

$$
M_z(t) = \sum_{k=-\infty}^{+\infty} M_{z(k)}e^{ik\Omega t}
$$

by expressing $S(\Omega)$ as a function of both positive and negative values of Ω and noting that $M_{z(+1)} = M^*_{z(-1)}$, in which case $S(\Omega) \propto \Omega|M_{z(+1)}|$. The emf induced in the pickup coil is then

$$
S(X) = S_0 X \left[\frac{1 + p^2\dfrac{X^2}{4}}{(1 + X^2)(1 + p^2 X^2)}\right]^{\frac{1}{2}}
$$

with the parameters defined as

$$\begin{vmatrix} X = \Omega T_1 \\ p = \dfrac{T_2}{T_1} \\ S_0 = aM_0 \dfrac{\mu_0 n Q_b}{RT_1} \dfrac{m}{(1+m^2)} \\ a = |\gamma|^2 H_1^2 T_2 T_1 \end{vmatrix}$$

Figure 11 depicts the variation of modulation signal $S(X)$ as a function of frequency for different values of ratio p. There are two distinct regions in this ensemble of curves that are determined by the relative magnitude of Ω and T_1. When $\Omega \ll T_1^{-1}$ (i.e.. $X \ll 1$), the signal in the coil reduces to $S(X) = S_0 X \propto \Omega$, that is, the spin system follows the modulation. The tangent at the origin, whose expression is independent of p, therefore represents the response of the system.

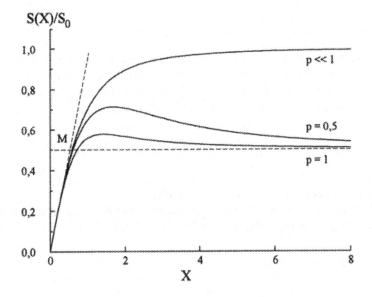

Figure 11. Variation of modulation signal $S(X=T_1)$, which is proportional to dM_z/dt, as a function of frequency for different values of $p = T_2/T_1$.

On the other hand, when $\Omega \gg T_1^{-1}$ ($X \gg 1$), the relaxation limits the exchange of energy between the spin system and the coil, and $S(X)$ tends to an asymptotic constant value equal to $S_0/2$. An approximation of the curve is then obtained by replacing it by the tangent at the origin together with its asymptote. The abscissa of the intersection point M relative to the ratio p is equal to

$$X = 1 \Rightarrow T_1 = \Omega^{-1}$$

when $p = 0$ $(T_1 \gg T_2)$, or

$$X = \frac{1}{2} \Rightarrow T_1 = \frac{\Omega^{-1}}{2}$$

when $p = 1$ $(T_1 = T_2)$. The profile of the curve therefore allows one to determine the spin–lattice relaxation time from the frequency at which the two lines constituting the profile intersect.

In order to measure ultra-short relaxation times less than 10^{-8} s, again the case $\Omega \ll T_1^{-1}$ applies. The curve $S(X)$ obtained in this case coincides with the tangent at the origin. One should then find the position of its asymptote. The modulation signal, $S(X)$, is independent of frequency Ω when it tends towards its asymptote, S_0 (value of the asymptote for $p \approx 0$). The cw-EPR signal, $\overline{S}$, is likewise constant, and it can be shown that the ratio

$$K = \frac{S_0}{\overline{S}}$$

is then independent of the characteristics of the sample. This property is used to determine ratio K with the help of a reference sample that is subjected to the same experimental conditions and chosen such that its T_1 time is greater than 10^{-8} s. With the help of the measured values of S_{0_r} and $\overline{S}_r$, one obtains

$$K = \frac{S_{0_r}}{\overline{S}_r}$$

where subscript r refers to the reference sample.

Ordinate S_0 of the asymptote of the material studied is thus determined from the knowledge of its cw-EPR signal, $\overline{S}_i$:

$$S_{0_i} = K \, \overline{S}_i = \frac{S_{0_r}}{\overline{S}_r} \overline{S}_i$$

And finally, the spin–lattice relaxation time to be determined is given by

$$T_1^{-1} = \frac{S_0}{\overline{S}_r} \frac{\overline{S}_i}{S_i(X)} \Omega$$

where

$$S_i(X) = S_{0_i} X = \frac{S_{0_r}}{\overline{S}_r} \overline{S}_i \, X .$$

4.2. Description of the Amplitude Modulation Spectrometer

The first version of an amplitude modulation spectrometer (Hervé & Pescia, 1960) has undergone many modifications (Pescia, 1965; Gourdon et al., 1973; Ablart & Pescia, 1980). It resembles a conventional cw-EPR spectrometer (Figure 12), except that a PIN-diode modulator and amplifier are inserted between the microwave source and the sample resonator. This arrangement enables one to modulate the microwave signal amplitude at frequencies ranging from 50 kHz to 30 MHz. An *in situ* pickup coil is tuned to angular frequency Ω and connected to a frequency-selective microvoltmeter.

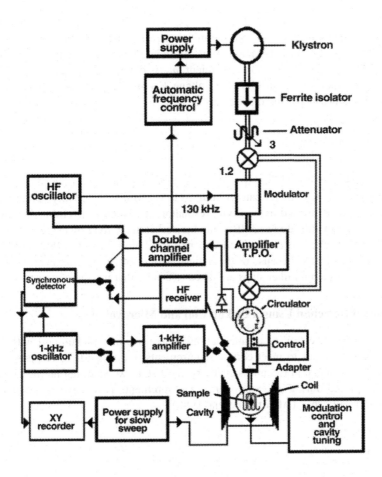

Figure 12. Block diagram of an amplitude modulation EPR spectrometer using a pickup coil and frequency-selective microvoltmeter as a detector.

The signal provided by the pickup coil is of the form $S(\Omega) = S_0 \times S(X)$. In order to measure spin–lattice relaxation times, it becomes necessary to ensure that S_0 is independent of Ω. This assumption, however, has not been verified experimentally. In fact, when Ω changes, the quality factor of the pickup coil (Q_b), the coefficient of modulation (m) and the amplitude of the radio-frequency field (H_1) all change, making it necessary to measure these parameters for each value of the frequency.

The ratio of the coefficient of modulation determined outside (m_e) and inside (m_i) the sample resonator cavity is

$$\frac{m_i}{m_e} = \left[1 + \left(\frac{2Q\Omega}{\omega}\right)^2\right]^{-1}$$

where parameters ω and Q denote the microwave frequency and resonator quality factor, respectively. Only the modulation coefficient inside the cavity varies with Ω and Q. The external modulation coefficient, m_e, is determined by using an antenna coupled to a matched load; effective voltage V_{eff} and DC voltage V_{DC} across the load are then measured, and one obtains

$$m(\%) = 100 \frac{V_{\text{eff}} \sqrt{2}}{V_{\text{DC}}}$$

The pickup coil is affixed to the sample tube using a saddle-shaped configuration (Figure 13). A Q-meter cannot be used to determine the quality factor of this device when it is inserted in the cavity resonator, and so Q_b is determined by measuring pass bandwidth $\Delta\Omega$ at −3 dB while the coil is tuned to Ω by using an external coil weakly coupled to a variable-frequency oscillator (Pescia, 1965); it follows that $Q_b = \Omega/\Delta\Omega$. Finally, because of the delicate nature of its measurement, H_1 is not directly measured but rather steps are taken to ensure that it remains constant.

4.3. Signal Detection Using a Pickup Coil and Microvoltmeter

The direct detection of magnetization via the use of a pickup coil within the sample resonator has to meet certain requirements with regards to response time, rejection of parasitic noise, and sensitivity. To begin, we shall define the range of Ω to be applied during an experimental measurement. This range of frequencies should optimally span (Ablart & Pescia, 1980)

$$\frac{1}{3} \frac{\Omega^*}{2\pi}$$

and

$$3 \frac{\Omega^*}{2\pi}$$

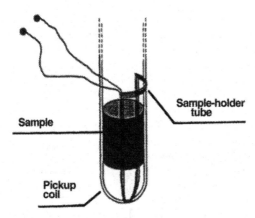

Figure 13. Saddle-shaped pickup coil used to detect magnetization changes in sample.

but the diode used for modulation and the pass bandwidth of the cavity in practice is confined to the range 10 kHz to 50 MHz. This range is going to determine the range of frequencies to which the detector must respond.

Second, the amplitude of signal collected by the pickup coil may reach 10 mV for those samples that contain a large number of magnetic moments, but for most materials the amplitude of the modulation signal is less than or equal to 10 μV. The sensitivity of the microvoltmeter employed in the modulation spectrometer (Figure 12) is approximately 0.5 μV. If it turns out to be sufficient to plot the asymptote of the modulation curve, the same may not be true for the tangent at the origin lying in the low-frequency region, where the background noise is large, making it diffi-cult to detect the signal. In practice, one can only measure relaxation times of those materials whose concentration of paramagnetic centers is more than 10^{15} centers per gauss.

In order to meet these demands of the amplitude modulation experiment, we have devised a new voltmeter whose schematic description appears in Figure 14. The signal is first amplified in a manner similar to that employed in a radio re-ceiver, after which the signal amplitude is measured by the use of synchronous detection. Data acquisition and control of this specialized detection device are made using a computer and RS-232 communication.

The strategy entails the use of synchronous detection with phase-locking. The underlying idea is to convert what is otherwise an AC signal into a DC signal, which enables one to use low-pass filters, rather than bandpass filters, to filter out the background noise. Low-pass filters can be constructed with passbands much smaller than bandpass filters, thereby providing a signal-to-noise enhancement of $4RC\Delta f$ for sinusoidal signals (Auvray, 1980). The input stage of the detector therefore consists of an impedance adapter followed by a wideband preamplifier and a mixer (Figure 15).

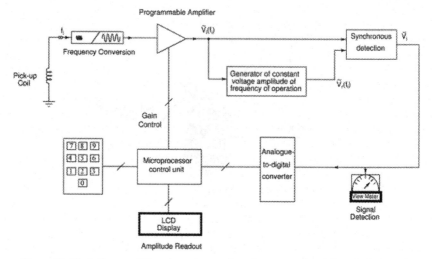

Figure 14. Block diagram of the microvoltmeter detection system, consisting of a frequency converter and a programmable amplifier. The processed signal may be monitored using an analog meter or LCD display during the setup and optimization stages of the experimental protocol.

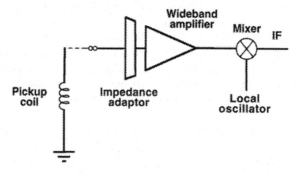

Figure 15. Schematic diagram of the frequency conversion stage of the microvoltmeter receiver, consisting of a wideband (pre)amplifier and mixer downconverter.

4.3.1. Stage 1: Preamplifier and Frequency (Down) conversion

The pre-amplifier (Figure 16) is constructed of discrete components, which has the advantage of producing an extremely satisfactory noise factor. A junction field-effect transistor, T1, serves as the input because it has a large gate impedance (10^{12} Ω) and small parasitic capacitance, which allows it to be used over a broad frequency range. The DC polarity of T1 is obtained by self-biasing via R3, R1, and inductance L1. With the transistor connected via a common source, one obtains a gain close to $-g_m R2$ since R3 is shunted via capacitor C1 (g_m is the transductance of the transistor and drain resistance is neglected; cf. Millman, 1979).

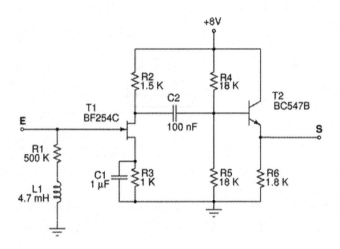

Figure 16. Schematic diagram of the wideband (pre)amplifier that is part of the frequency (down)converter stage of the microvoltmeter detector (cf. Figure 15).

The output impedance of T1 (equal to R2) is much higher than the 1.5-kΩ input impedance of the frequency-conversion stage, and so bipolar transistor T2 is inserted in the emitter-follower (common collector) configuration. The voltage gain of T2 is close to unity, and its output impedance is equal to a few ohms ($\beta =$ 100; cf. Millman, 1979). So configured, the amplifier stage behaves as a voltage source.

The frequency response of the preamplifier is illustrated in Figure 17. This curve, plotted on a semi-logarithmic scale, shows the variation of the output signal, S, when a signal, E, of variable frequency and constant amplitude, is applied at the input. It is noted that the gain is constant between 5 kHz and 10 MHz, which covers the entire range of frequencies employed to measure relaxation times. Furthermore, it is independent of the impedance of the pickup coil.

After preamplification the signal is up/down converted via a two-step superheterodyne process. The first stage converts the amplified input signal to a 10.7-MHz signal that is filtered and passed to a second mixer that converts the first stage signal down to the desired 70-kHz intermediate frequency (Figure 18). The first stage is therefore tuned so as to ensure that beat frequency $f_0 + f_{loc}$ is equal to 10.7 MHz.

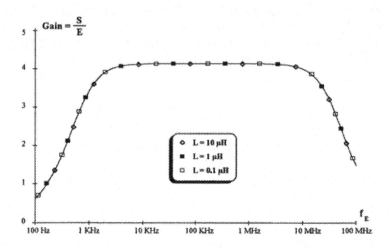

Figure 17. Variation of the preamplifier gain as a function of the input signal frequency for different values of the pickup coil inductance.

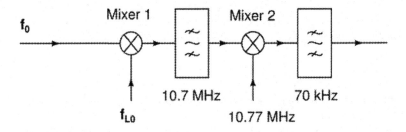

Figure 18. Mixer portion of the frequency converter stage of the microvoltmeter detector. Circuit features a two-step conversion from f_0 to 70 kHz.

The rationale for using the two-stage conversion process may be understood by examining the selectivity of the one- *vs.* two-mixer configuration. In the former case, a single mixer is used with a variable local oscillator and a 70-kHz bandpass filter to the convert amplitude modulation signal (frequency between 100 kHz to 10 MHz) to the IF of 70 kHz. The maximum and minimum frequencies passed by the 70 kHz filter are denoted by

$$f_{70M} = f_{70} + \frac{\Delta f_{70}}{2}$$

and

$$f_{70m} = f_{70} - \frac{\Delta f_{70}}{2}$$

respectively. Parameters f_{70} and Δf_{70} denote the center frequency and the passband (the latter given by the ± 3 dB points). The requisite local oscillator frequencies that will attain f_{70M} and f_{70m} are

$$f_{70_M} = f_{loc_M} - f_0$$

and

$$f_{70_m} = f_{loc_m} - f_0$$

which means that the allowable dispersion of frequencies in the local oscillator is restricted to

$$f_{loc_M} - f_{loc_m} = \Delta f_{loc} = \Delta f_{70} = f_{70_M} - f_{70_m}$$

or, in other words, its quality factor corresponds to

$$Q_1 = \frac{f_{loc}}{\Delta f_{loc}} = \frac{f_{loc}}{\Delta f_{70}}$$

It is now necessary to determine how the selectivity of the detector will change as Ω is varied. Since the frequency of the local oscillator must equal $f_0 + f_{70}$ (f_0 equivalent to Ω) the quality factor Q_1 is expressed as

$$Q_1 = \frac{f_0}{f_{70}} \frac{f_{70}}{\Delta f_{70}} + \frac{f_{70}}{\Delta f_{70}} = Q_{70} \left(1 + \frac{f_0}{f_{70}} \right)$$

where Q_{70} is now the quality factor of the bandpass filter. This last expression shows that the frequency selectivity of the detector depends strongly on Ω, which in practice will always be greater than 70 kHz.

In the case of a double mixer converter, the maximum and minimum frequencies at the output stage are modified:

$$f_{70M} = 10.77 \text{ MHz} + f_{10.7m}$$

and

$$f_{70m} = 10.77 \text{ MHz} - f_{10.7m}$$

Since $f_{10.7M} = f_{locM} - f_0$ and $f_{10.7m} = f_{locm} - f_0$, one finds that $\Delta f_{loc} = \Delta f_{10.7} = \Delta f_{70}$. And since the pass bandwidth of the filter centered at frequency 70 kHz is narrower than that of the filter centered at 10.7 MHz, one obtains again the relation

$\Delta f_{\text{loc}} = \Delta f_{70}$. Thus, taking into account the fact that $f_{70} = 10.77$ MHz$-(f_{\text{loc}}-f_0)$, the total selectivity of the ensemble of two mixers becomes

$$Q_2 = \frac{10.77 \text{ MHz} + f_0}{f_{70}} \frac{f_{70}}{\Delta f_{70}} \frac{f_{70}}{\Delta f_{70}} = Q_{70}\left(\frac{10.77 \text{ MHz} + f_0}{f_{70}} - 1\right)$$

Quality factor Q_2 is much greater than quality factor $Q1$ and varies insignificantly with the frequency of the input signal.

This analysis demonstrates clearly that two-stage frequency conversion enables us to increase the selectivity of our apparatus. It also makes it possible for us to reduce the f^{-1} and f^{-2} noise with the help of a very selective filter (*e.g.*, a ceramic filter centered at 10.7 MHz) inserted in between the two mixers. Figure 19 shows the distribution of different frequencies available in the frequency domain at the inputs and outputs of a double-mixer circuit. In order that the intervals do not overlap each other, an operating range of frequencies 500 kHz to 9 MHz has been chosen here.

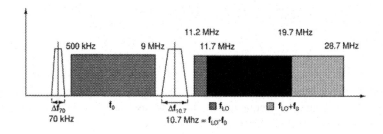

Figure 19. Representation of the frequency distribution of the double mixer converter circuit.

4.3.2. Stage 2: The Programmable Amplifier

The next stage of the microvoltmeter detector consists of a programmable variable gain amplifier that amplifies the 70-kHz IF signal prior to filtering so as to compensate for signal amplitude losses incurred during the mixer stages. This amplifier is connected to the output of the filter (Figure 20). The 70-kHz IF filter consists of a series RLC circuit (passive) because it offers certain advantages over an operational amplifier (active) filter; among these advantages are ease of construction and a satisfactory noise factor. Details of the IF filter and its associated amplifiers are shown in Figure 21. The signal exiting the second mixer stage arrives at resistor R1 across capacitor C1 and is amplified by the first operational amplifier whose gain is equal to 14 because at resonance R_L forms a voltage divider bridge along with the output impedance of the first op-amp. The current that passes the series RLC filter is converted into a voltage by the second operational amplifier,

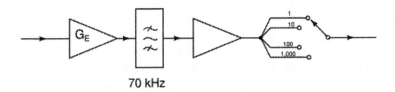

70 kHz

Figure 20. Overview of the programmable decade amplifier module, consisting of a variable gain input amplifier, a 70-kHz filter, and an output amplifier with stepwise adjustable gain.

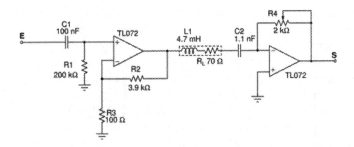

Figure 21. Schematic diagram of the 70-kHz RLC filter.

the amplitude of which is controlled via R4. Taking into account the properties of the operational amplifier, the current that passes series RLC filter is

$$i = \frac{E'}{R_L + j\left(L\omega - \dfrac{1}{C\omega}\right)}$$

whereas the voltage output of the operational amplifier (*i.e.*, signal S) is $S = -R_4 i$. Voltage E' is thus amplified with the gain

$$G = \frac{S}{E'} = -\frac{R_4}{R_L} \frac{1}{\sqrt{1 + \dfrac{1}{R_L^2}\left(L\omega - \dfrac{1}{C\omega}\right)^2}}$$

At resonance, the gain of the filter is equal to the ratio of the resistance values and its quality factor $Q = (L\omega_c / R_L)$ is independent of R4. One can thus modify the gain without changing the pass bandwidth, which, moreover, depends only on internal resistance R_L of the inductance itself.

It remains to be determined whether the gain should be controlled by resistance R4 or R2. Figures 22 and 23 depict simulations of the gain and signal-to-noise ratio as functions of frequency when R4 or R2 are varied, respectively. Figure 22 shows that the variation of the gain of the filter is proportional to the value of resistance R4. Furthermore, the signal-to-noise ratio is practically constant at the resonant frequency. On the other hand, it is noted that there appears a slight modification of this frequency, mainly for large values of R4. If the gain is controlled via R2, there no longer exists a linear relation between the gain and the resistance (Figure 23). This also affects the signal-to-noise ratio, whose variations are more significant than those in the preceding case. In addition, one should also note that there is a clear-cut shift in the resonant frequency. It therefore follows that the gain of the filter should be controlled by resistance R4. The values of the components used in the circuit lead to a pass bandwidth of about 4 kHz.

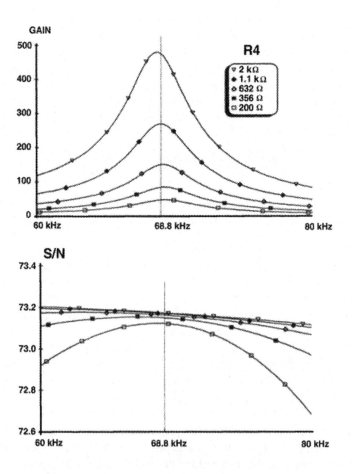

Figure 22. Variation of gain and S/N ratio as R4 (*cf.* Figure 21) is varied.

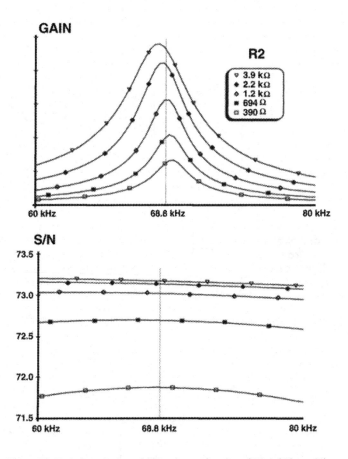

Figure 23. Variation of gain and S/N ratio as a function of R2 (*cf.* Figure 21).

At the intended frequency of operation, modern operational amplifiers cannot be used at a gain greater than a few multiples of 10. The amplification needs of the detector are therefore met by connecting three identical amplifiers in series, with each stage programmed to pass the signal with gain of unity or 10 (Figure 24). It is important, however, that the amplifiers and associated circuit components be carefully matched so as not to propagate errors.

The structure of a single amplifier stage is illustrated in Figure 25. An amplifier is connected in the non-inverter amplifier mode, in which the gain is equal to 1 + R2/R1. A NAND gate is used to provide two-state amplifier control via a relay that switches the feedback loop of the operational amplifier from follower with unity gain to the amplifier with gain 10, as set by R2. The advantage of using the relay, as opposed to a solid-state switch, is that it provides a true zero resistance when closed and complete DC isolation between the digital and analog circuitry.

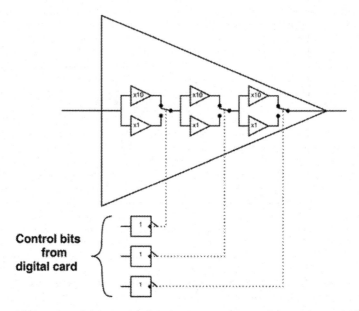

Figure 24. Overview of the stepwise (decade) programmable amplifier consisting of three sequential units.

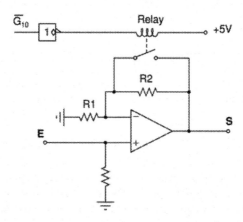

Figure 25. Schematic diagram of a single amplifier unit from the decade programmable amplifier depicted in Figure 24.

4.3.3. Stage 3: Synchronous Detection of the Signal

The frequency downconverter and amplifier stages ensure that the signal amplitude coming from the pickup coil is at least 100 mV. But despite the narrow bandwidth of the filters, the level of noise that passes receiver Stages 1 and 2 is still too high. The third and final stage of the microvoltmeter detector is therefore dedicated to signal reception and further noise reduction, as well as data recovery and display. These functions are achieved by using a synchronous detection circuit together with a liquid-crystal display for reading and a view meter for the detection of the signal. These are depicted as a block diagram in Figure 26.

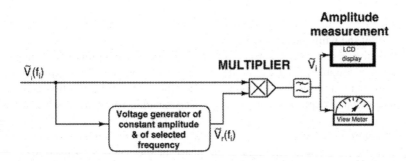

Figure 26. Block diagram of the synchronous detector and data output stage of the microvoltmeter detector. The signal obtained from the programmable amplifier (stage 2) is used to create a stable reference signal via a phase-locked loop, which is multiplied against the signal in order to obtain the DC amplitude.

The signal, S_i, that is obtained at the output of the programmable-decades amplifier may be considered as the sum of the signal whose amplitude is to be measured and the voltage of the noise transmitted:

$$S_i = A_i \cos(\omega_s t) + e(t)$$

where $A_i \cos(\omega_s t)$ is the desired information, and $e(t)$ corresponds to the noise.

As long as amplitude A_i is larger than the noise voltage, $e(t)$, the signal remains distinct and can be measured. But if the two voltages are close to each other, the noise voltage completely masks the desired information, and only a filter possessing an infinite value of the quality factor and centered at frequency ω_s will enable determination of amplitude A_i.

Supposing now that one supplies a reference signal S_r of the same angular frequency as the desired signal, that is, ω_s, and possesses known amplitude A_r: $S_r = A_r \cos(\omega_s t + \varphi)$. By multiplying the two voltages, one obtains a signal equal to

$$S_i \times S_r = [A_i \cos(\omega_s t) + e(t)] A_r \cos(\omega_s t + \varphi)$$

whose spectral decomposition can be expressed as follows:

$$S_i \times S_r = \frac{A_i A_r}{2} [\cos(\varphi) + \cos(2\omega_s + \varphi)] + e(t) \times S_r$$

One thus obtains a term corresponding to a DC voltage:

$$\frac{A_i A_r}{2} \cos(\varphi)$$

a term of double frequency:

$$\frac{A_i A_r}{2} \cos(2\omega_s t + \varphi)$$

and a term for the white noise: $e(t) \times S_r$.

The quality factor of a low-pass filter is much greater than that of a bandpass filter, and so a low-pass filter can be used to eliminate the noise by recovering the desired information in the DC component. Since the value of the amplitude of signal S_r is known and dephase angle φ can be determined, the unknown value of amplitude A_i of the signal can be directly obtained. This principle is familiar during the operation of a cw-EPR spectrometer, in which the receiver is tuned to the field modulation frequency and one adjusts the phase of the receiver to maximize the receiver voltage.

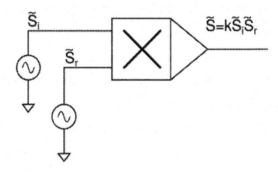

Figure 27. Four-terminal voltage multiplier, which provides synchronous detection of signal amplitude.

The product is accomplished with the help of a four-terminal multiplier (Figure 27). This circuit produces a voltage, $\tilde{S} = k\, \tilde{S}_i \tilde{S}_r$, regardless of the signs of $\tilde{S}_i$ and $\tilde{S}_r$. Factor k allows one to adjust the amplitude of the output voltage. In a practical synchronous detector one should add two circuits to the multiplier. The first is a low-pass filter to separate DC from AC voltage. This filter is simply an RC combination whose time constant is equal to 300 ms, which represents a good compromise between high quality factor and rapid response time. The second addition is a phase corrector that allows one to adjust factor $\cos(\varphi)$ so as to allow one to maximize the output voltage.

For our purposes, the frequency that is used to modulate H_1 is not suitable for synchronous detection because of the wide range of prospective frequencies and the associated technical demands of constructing a phase corrector. As an alternative, the reference signal is derived from the signal delivered at the output of the programmable decade amplifier (stage 2) via a phase-locked loop (PLL) circuit, as depicted in Figure 28. This PLL circuit consists of a sine-to-square wave converter, a frequency divider, and a conventional PLL that controls a digital voltage-controlled oscillator (VCO). The stabilized output of this VCO serves as S_r at the multiplier input depicted in Figure 27.

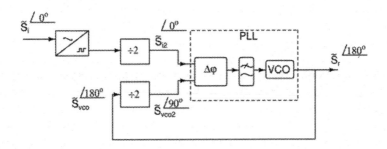

Figure 28. Block diagram of the circuit used to generate a stable reference signal from the IF signal coming from stage 2 (corresponds to the voltage generator block depicted in Figure 26). A portion of this signal, S_i, is converted to a square wave, has its frequency divided by 2, and is then phase-locked to a VCO also running at the IF.

The use of a PLL to provide a stable reference signal for synchronous detection has certain advantages for our application. The voltage-controlled oscillator is set to operate at the intermediate frequency of the detector stages 1 and 2, that is, 70 kHz. The operating range of the VCO, namely f_{min} and f_{max}, provides noise rejection because any frequency outside of this range will be ignored by the PLL. It therefore stands to reason that one desires to minimize the operating range of the VCO (and hence the PLL).

The digital VCO generates a square-wave TTL signal whose frequency can be adjusted by a continuous voltage (0 to 5 volts). We specify the center (operating range) frequency f_c control voltage as $\frac{1}{2}V_{cc}$, and the locking range as $2F_L$. A capacitor and two resistances fix the VCO frequency limits, and phase locking is achieved by a circuit that consists of three phase comparators (Philips Application Notes for the PLL 74HC/HCT4046A, p. 721 (1986)). These comparators are of the XOR *(eXclusive OR)* type and are coupled to a low-pass passive filter (Figure 29). The amplitude of the phase comparator $S_{\Delta\varphi}$ is adjusted so as to match the VCO operating control range (0 to V_{cc}).

The signal derived from the stage 2 output is sinusoidal, and that portion which is used to create S_r must therefore be converted to a square-wave signal in order to be compatible with and locked to the VCO via the comparators described in the preceding paragraph. This is achieved using the circuit depicted in Figure 30; the gain pass-bandwidth of the amplifier is 500 MHz and the slew rate is 500 $V \cdot \mu s^{-1}$. Provision for controlling the offset voltage enables one to compensate for dephasing between the input and output signals.

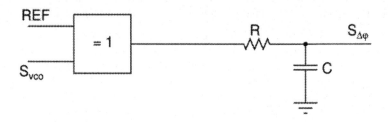

Figure 29. Electrical schematic of a phase comparator: continuous voltage $S_{\Delta\phi}$ is proportional to the phase difference between REF and S_{VCO}.

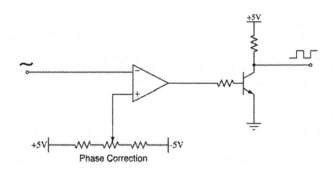

Figure 30. Circuit for conversion of a sinusoidal signal of amplitude −5 to +5 V into a square wave signal of 0 to +5 V with phase correction.

The frequency dividers shown in Figure 28 are necessary in order to ensure that the VCO output is nonzero. To create the reference signal for the synchronous detector, one begins by setting the VCO frequency to the intermediate frequency of stages 1 and 2. During signal detection, the output voltage from the filter (mixer circuit, Figure 18) grows during the period over which one approaches the frequency of concurrence. The VCO synchronizes itself for those frequencies within its operating range and phase angle 0 to 180°. Under these conditions the voltage supplied by the synchronous detection circuit would be zero. But this latter problem is corrected by introducing the frequency dividers at each input of the phase comparator (*cf.* Figure 31), which ensures that the dephasing angle between the signal and the VCO is always 180°.

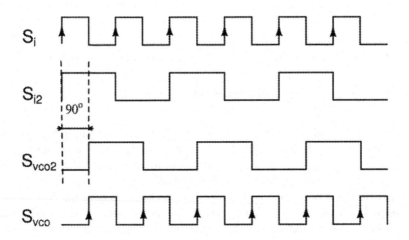

Figure 31. Chronogram of the signals present in the circuit that is used to generate the reference signal.

The inherent noise discrimination by the VCO, which was cited in a preceding paragraph, may be understood from the so-called cyclic overlap fraction of the PLL. With both the input (*i.e.*, the now modified S_i) and feedback signals of the loop taken to be square waves, a representative chronogram of the phase comparator component of the PLL is shown in Figure 32. The output is a periodic signal

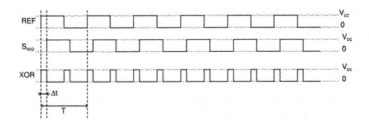

Figure 32. Variation of the output voltage of an XOR gate for two out-of-phase signals of the same frequency.

whose frequency is twice that of the VCO. This sequence of pulses is converted to a square-wave signal that depends on the cyclic-overlap fraction, k, which is defined as the ratio of the time interval when the signal is nonzero and the half-period of the signal. It is represented by function

$$\begin{cases} f(t) = V_{cc} \; ; \; 0 \le t < kT \\ f(t) = 0 \; ; \; kT \le t < T \end{cases}$$

where T denotes the period of the signal.

Only the DC component, kV_{cc}, of the Fourier series expansion, that is,

$$f(t) = kV_{cc} + \sum_{n=1}^{\infty} \frac{4V_{cc}}{n\pi} \sin(nk\pi) \cos(n\omega\, t)$$

contains the information on the cyclic-overlap fraction, k, which, in turn, provides information on the dephase angle between the two signals. The nonzero components of the frequency are then eliminated from the signal with the help of a low-pass filter in order to retain only the useful voltage.

Let Δt be the time interval that separates the beginnings of the periods of the two signals (Figure 32). The dephase angle, $\Delta\varphi$, is therefore equal to $2\pi^{\Delta t}/_T$, where T is the period of the reference. The voltage at the output of the XOR gate is at a high level during interval Δt. From the definition of the cyclic-overlap fraction, one therefore deduces:

$$k = 2\frac{\Delta t}{T}$$

(The factor 2 appears because the frequency of the signal at the XOR gate is twice that of the VCO signal).

By eliminating term Δt between k and $\Delta\varphi$, one obtains

$$k = \frac{\Delta\varphi}{\pi}$$

The amplitude of the DC (feedback) voltage, $S_{\Delta\varphi}$, that determines the frequency, f_{vco}, of the oscillator is equal to

$$kV_{cc} = \frac{\Delta\varphi}{\pi} V_{cc}$$

From the voltage-frequency transfer curve, one obtains the expression for the frequency as a function of the dephase angle:

$$f_{vco} = \frac{2}{\pi} \Delta\varphi (f_c - f_{min}) + f_{min}$$

It is important to note that a nonzero dephase angle between the VCO and refer-ence remains constant when their frequencies are the same. In fact, if one expresses the phase difference as a function of the frequency of VCO

$$\Delta\varphi = \frac{\pi}{2}\frac{f_{vco} - f_m}{f_c - f_m}$$

one finds that to each frequency there corresponds a unique value of the dephase angle. In the particular case when $f_{vco} = f_c$, the phase difference is equal to $\pi/2$.

When the reference signal is absent, the XOR gate reproduces the signal of the VCO. The phase comparator then provides voltage

$$S_{\Delta\varphi} = \frac{V_{cc}}{2}\left(k = \frac{1}{2}\right)$$

The frequency of the VCO that corresponds to this voltage is the central frequency, f_c, and so it follows that this center frequency corresponds to the free-running fre-quency of the VCO.

4.3.4. Signal Detection and Display

At the onset of the experiment, one tunes the microvoltmeter to the signal whose frequency corresponds to the sum of the frequency of the coil and 10.7 MHz ($f_{local} = f_{coil} + 10.7 \times 10^6$). The high degree of selectivity of the mixer circuit makes it difficult to detect this equality by a liquid-crystal display (LCD), and so the DC voltage supplied by the multiplier circuit is directed partly to an analog-to-digital converter and partly to a galvanometer with a needle (Figure 33). One thereby de-termines by visual inspection of the galvanometer the moment when the micro-voltmeter is tuned to the correct frequency. In order to fine-tune the control of the frequency of concurrence, one can also control the sensitivity of the view meter, regardless of the value of the gain programmed on the decade amplifier. Once one has obtained a maximum deviation of the galvanometer needle, the effective value of the amplitude of the signal may be read on the LCD display.

The low-pass filter associated with the multiplier consists of an RC unit with the time constant equal to 300 ms. Two necessary conditions dictate the choice of the time constant. First, a very long duration results in a long reaction time of the galvanometer needle, which may lead to the risk of not being able to detect the concurrence. Second, too short a duration does not allow filtering out effectively the radio-frequency parasites, at which point the galvanometer needle becomes unstable and vibrates excessively.

Conversion of the DC voltage supplied by the synchronous detector to a digital value is effected by a card built on the basis of an 8-bit MC6809 microprocessor. The A/D converter circuit also consists of a ROM that contains a control program (in assembler), a RAM for execution of the program and data storage, a PLA inter-face to eliminate "fan-out" problems, an A/D converter, and an RS232 inter-

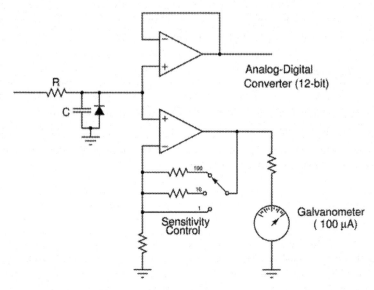

Figure 33. Circuit for detection of the signal induced in the measuring coil (galvanometer) and its direct reading via an A/D converter.

face to the computer. The converter used requiring the use of an input voltage between 0 and 10 V possesses a resolution of 12 bits. In this manner one obtains a display of 4096 different values with an interval of 2.44 mV between successive points. This allows a reading of the amplitude with a precision better than 1%.

The assembler language is subdivided into five main parts: keyboard management, display of voltage reading (on the average of 50 values), control of the gain of the analog card, management of the RS232 connector. and, finally, test of certain parts (screen, keyboard, ADC).

4.4. Performance Tests of the Microvoltmeter Detector

The detector is protected from external noise by enclosing it within a grounded metallic box, and power is supplied via a local filter. Those circuits that are sensitive to electronic noise (mixers and preamplifiers) were isolated with the help of aluminum cases connected to the analog ground of the electronic assembly. The assembled selective microvoltmeter assembled detector was evaluated for its sensitivity, linearity, and pass bandwidth. This study was carried out with the help of a calibrated HF generator that was attached to a calibrated attenuator and so provided a sinusoidal voltage of amplitude 0.1–500 µV and a frequency tunable between 0.5 and 8 MHz.

At signal amplitudes below 1 µV, large fluctuations prevent a correct reading of the signal amplitude. Nevertheless, the galvanometer of the microvoltmeter can detect a signal as small as 0.1 μV. The linearity of a system expresses the conformity of transformation between the input and output signals. In the case of our voltmeter, it allows to determine the exact level at which the voltage gain remains constant regardless of the amplitude of the modulation signal. A signal of frequency 1 MHz was used to measure the voltage gain of our apparatus. It remains proportional (Figure 34) to the input voltage (at ±1 dB) when its value lies between 1 and 500 μV. (The performance is better than this at higher frequencies. where there exists less electronic noise.)

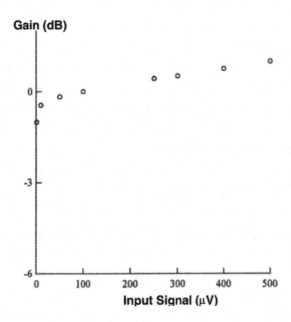

Figure 34. Variation of the amplifier gain with signal amplitude.

The pass bandwidth allows one to determine the effective amount of the signal that will pass on to the mixers. It is necessary that it be wider in frequency than the range of frequencies of modulation signals, but it is also necessary that it does not allow the superfluous noise to pass through, especially, the background noise which is most important. In order to determine the pass bandwidth at −3 dB of our microvoltmeter, a voltage of 500 μV was applied whose frequency was varied between 0.5 and 8 MHz; the variation of the amplitude measured was 1.1±0.2 dB over the entire frequency spectrum (Figure 35), and so the cutoff frequencies lie outside of this frequency interval.

In summary, the performance of our apparatus enables measurement of the amplitude of a signal with a precision of 2 dB between 1 and 500 μV over a pass bandwidth larger than 8 MHz. The measurements carried out on a calibrated sample (Varian pitch) permitted determination of new performances of the modulation spectrometer. In particular, the improvement of the signal-to-noise ratio allowed an increase in the sensitivity by a factor of 10, that is, 10^{14} centers per Gauss.

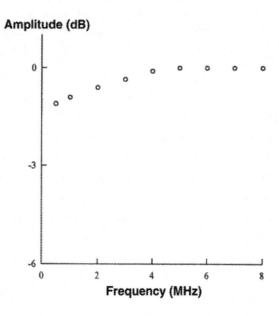

Figure 35. Variation of the amplitude of the mixers as a function of the input signal frequency.

5. CONCLUSIONS

In this chapter we have dealt with the problem of measuring spin–lattice relaxation time from the very short to the very long using a modulation spectrometer. The measurement of spin–lattice relaxation times less than 1 µs is best performed by modulating the amplitude of the incident microwave field and following the magnetization of the spin system via a pickup coil surrounding the sample material. As described in §4 of this chapter, detection of M_z was optimized by modifying the receiver that is used with the pickup coil. This receiver measures the emf generated within the coil by the time-variant (*i.e.*, modulated) magnetization, and it consists of a selective microvoltmeter that operates in the frequency range 0.2 to 9 MHz. The broad operating (frequency) range of the microvoltmeter is made possible by using a "superheterodyne" scheme with double-mixing of frequency associated with synchronous detection of the amplitude of the signal to be measured, achieved by a phase-locking loop. The ensemble of electronic circuits is controlled by a microprocessor that permits a choice of sensitivity, a visual display of the signal amplitude, and communication between the spectrometer and a computer.

The synchronous microvoltmeter detector allows linear (to ±1 dB) measurements of signal amplitudes of 1 to 500 µV with a band passing at −3 dB higher at 9 MHz. Between 0.1 and 1 µV, significant noise prevents a precise reading of the amplitude, which however remains proportional to the signal to be measured, contrary to the situation with the unmodified spectrometer.

On the other hand, we do not have available a simple efficient technique adapted to measure T_1 times greater than 1 µs in our laboratory, apart from the method of continuous saturation, which possesses the disadvantages of sample heating, difficulty of measuring H_1^2, and requisite measurement of T_2. For these measurements we therefore developed an original technique that is based on Bloch's phenomenological equations, whose solution by Laplace transform allows one to derive a particular form of the solution at resonance. This solution is a transfer function of second order, with a quality factor of greater than unity and a cutoff frequency equal to $(2T_1)^{-1}$ at −3 dB, and this approach to data analysis allows one to design a method based on sweeping static field H_0 with a small amplitude compared to the linewidth. The variation of the modulus of transfer function so obtained allows one to determine spin–lattice relaxation times longer than 3 µs.

A practical test of the spectrometer was performed on two sample materials, for which we examined the temperature dependence of the spin–lattice relaxation time, that is, $T_1^{-1}(\Theta)$. In the first case study, we examined the T_1 behavior of calcium cadmium acetate tetrahydrate doped with 5% copper. This sample offered the advantage of providing a relatively strong signal despite an extremely weak exchange interaction ($J_N \sim 10^{-2}$ K), and the crystal undergoes a phase transition at a temperature that depends on the concentration of Cu^{+2} ions (Shields et al., 1992). At high temperatures (140 K < Θ < 300 K) T_1^{-1} varies linearly with respect to Θ^2, which indicates that the spin–lattice relaxation proceeds by a Raman process (van Vleck, 1940). We were facing particularly favorable circumstances here to observe the relationship $T_1^{-1} \propto \Theta^2$ taking into account an exchange interaction quasi-nonexistent and the opportunity of having at our disposal a technique well adapted to measuring very short T_1 times. At lower temperatures (70 K < Θ < 120 K), the observed temperature dependence of T_1^{-1} varying as Θ^4 indicated that the sample temperature is lower than the Debye temperature (the dependence T_1^{-1} varying as Θ^9 could not be achieved since it would require a temperature lower than 2 K, where one would observe a Θ^n dependence, $2 < n < 9$).

From our data we deduced the Debye temperature of 135 K for the crystal, in good agreement with the value of 148 K found by Nusawa et al. (1975). At yet lower temperatures (less than 70 K), the curve indicates coexistence of two opposing effects: the Raman process and occurrence of antiferromagnetic ordering with a Néel temperature close to absolute zero. As a consequence, we could not confirm the influence of phase transition on relaxation mechanisms.

A second case study was performed on a borate glass doped with 0.25% of Fe_2O_3. At lower temperatures (less than 130 K), the behavior of the curve $T_1^{-1}(\Theta)$ ($T_1^{-1} \propto \Theta^2$ below 30 K and $T_1^{-1} \propto \Theta$ above 30 K) is in good agreement with the model of Kurt & Stapleton (1980) proposing modulation of the hyperfine interaction by phonons. At higher temperatures (greater than 130 K), the asymptotic character of the curve $T_1^{-1}(\Theta)$ suggests invoking the three-reservoir model (Zeeman, exchange, lattice) of Bloembergen & Wang (1954), but we could not determine the exact nature of the intervening reservoir. The result obtained for this glass differs entirely from that reported for silicate glasses by Θ (Bouchina *et al.* (1991).

In the quest for optimizing the modulation spectrometer one could attempt to change the intermediate frequency of the "superhetrodyne" receiver of our microvoltmeter (455 kHz instead of 70 kHz). One could also replace the present cavity by a dielectric cavity that will support the presence of the pickup coil better, and render automatic the process of acquisition of data.

6. LITERATURE CITED

Abragam A. 1961. *Principles of nuclear magnetism.* Oxford: Clarendon Press.

Abragam A, Bleaney B. 1970. *Electron paramagnetic resonance of transition ions.* Oxford: Clarendon Press.

Ablart G. 1978. Étude de la relaxation spin réseau de certains sels ioniques à partir des dépendences $T_1^{-1}(T)$, Thèse d'état, Université Paul Sabatier, Toulouse.

Ablart G, Pescia J. 1980. *Phys Rev B* **22**:1150.

Alger RS. 1968. *Electron paramagnetic resonance: techniques and applications.* New York: Wiley Interscience.

Anderson PW, Weiss PR. 1953. *Rev Mod Phys* **25**:269.

Auvray J. 1980. *Electronique des signaux analogiques.* Dunod Université.

Badoual R. 1984. *Les micros-ondes: I — circuits. microrubants. fibres.* Paris: Masson.

Basso C. 1990. Problèmes de masse et techniques de découplage en hautes et basses fréquences. *Radio Plan*, Mai, no. 510, p. 59.

Bloch F. 1946. *Phys Rev* **70**:460.

Bloembergen N, Purcell EM, Pound RV. 1948. *Phys Rev* **73**:679.

Bloembergen N, Wang S. 1954. *Phys Rev* **93**:72.

Bowers KD, Mims WB. 1959. *Phys Rev* **115**:285.

Boucher JP, Nechtschein M. 1970. *J Phys* **31**:783.

Bourdel D. 1980. Relaxations des spins electroniques dans un isolant magnétique unidimensionnel: le TMMC evolution des temps T_1 et T_2 avec l'orientation et la température. Thèse d'état, Université Paul Sabatier, Toulouse.

Bouhacina T, Ablart G, Pescia J. 1991. *Solid State Comm* **78**:573.

Buishvili LL, Khalvashi KH. 1973. *Phys Stat Solidi C* **55**:13.

Encinas J. 1989. *Systemes à verrouillage de phase (P.L.L.).* Paris: Masson.

Feher G. 1957. *Bell Syst Techn J* **36**:449.

François P, Catala A, Scouarnec Ch. 1993. *Med Phys* **20**:1695.

Gill JC. 1962. *Proc Phys Soc* **79**:58.

Goldsborough JP, Mandel M, Pake GE. 1960. *Phys Rev Lett* **4**:13.

Gorter CJ. 1947. *Paramagnetic relaxation.* London: Elsevier.

Gourdon J-C. 1975. De l'égalite des temps T_1 et T_2 dans la resonance electoniques des meataux purs et des alliages dilues. Thèse d'état, Université Paul Sabatier, Toulouse.

Gourdon J-C, Lopez P, Pescia J. 1968. *Peintures, Pigments, Vernis* **44**:257.

Gourdon J-C, Vigouroux B, Pescia J. 1973. *Phys Lett A* **45**:69.

Gregson AK, Mitra S. 1969. *J Chem Phys* **50**:2021.

Grivand P, Blaquière A. 1958. *Cours d'electronique: le bruit de fond*, Vol. 4, pp. 1–19, 128–144. Paris: Masson et Cie.

Hahn EL. 1950. *Phys Rev* **80**:580.

Halbach K. 1954. *Helv Phys Acta* **27**:259.

Hervé J, Pescia J. 1960a. *CR Acad Sci* **251**:665.

Hervé J, Pescia J. 1960b. *Arch Sci Fasc Spec* 350.

Johnson JB. 1928. *Phys Rev* **32**:97.

Kiel A, Mims WB. 1967. *Phys Rev* **161**:386.

Kronig R. 1939. *Physica* **6**:33.

Kurtz SR, Stapleton MJ. 1980. *Phys Rev B* **22**:2195.

Lagasse J. 1964. *Andude des circuits electriques*. Paris: Eyrolles.

Langs DA, Hare CR. 1967. *Chem Commun* 890.

Larson GH, Jeffries CD. 1966. *Phys Rev* **141**:161.

Lees RA. 1967. PhD dissertation, Nottinghan University

Locker DR, Look DC. 1968. *J Appl Phys* **39**:6119.

Look DC, Locker DR. 1968. *Phys Rev Lett* **20**:987.

Lopez P, Gourdon J-C, Pescia J. 1968. *Peintures-Pigments-Vernis*, **44**:496.

Lopez P, Gourdon J-C, Pescia J. 1971, *CR Acad Sci, Paris*, Ser. B, **273**:239.

Millman I. 1979. *Microelectronics: digital and analog circuits and systems*. New York: McGraw-Hill.

Mims WB. 1965. *Rev Sci Instrum* **36**:1472.

Misra SK, Kumar K. 1986. *J Chem Phys* **84**:2514.

Misra SK, Kahrizi M. Korczak S. 1992. *Physica B* **182**:186.

Moriya T. 1960. *Phys Rev* **120**:91.

Morocha AK. 1963. *Sov Phys Sol Stat* **4**:1683.

Nogatchewsky M, Ablart G, Pescia J. 1977. *Sol. Stat. Com.* **24**:493

Nyquist H. 1928. *Phys Rev* **32**:110.

Orbach R. 1961a. *Proc Roy Soc A* **264**:458.

Orbach R. 1961b. *Proc Roy Soc A* **264**:485.

Orton IW. 1968. *Electron paramagnetic resonance*. London: ILIFFE Books.

Pake GE. 1962. *Paramagnetic resonance: an introductory monograph*. New York: W.A. Benjamin.

Pescia J. 1965. La mesure des temps de relaxation spin–réseau trés courts. *Ann Phys* (Paris) **10**:389.

Pescia J. 1966. *J Phys* **27**:782.

Pescia J, Hervé I. 1964. Proc. XIIth CoIl. Ampere, Bordeaux 17–21 Sept 1963. Amsterdam: North-Holland, p. 247.

Poenadji S. 1985. Dynamique de spins electroniques dans un système unidimensionnel à couplage ferromagnétique le C.H.A.C. Thèse d'état, Université Paul Sabatier, Toulouse.

Pouw CLM, Van Duyneveldt AI. 1976. *Physica* **81b**:15.

Scott PL, Jeffries CD. 1962. *Phys Rev* **127**:32.

Shields H, Kleman TG, De DK. 1992. *J Chem Phys* **97**:482.

Singer LS, Kommandeur J. 1961. *J Chem Phys* **34**:133.

Slichter CP, Purcell EM. 1949. *Phys Rev* **76**:466.

Standley KJ, Wright J. 1964. *Proc Phys Soc* **83**:361.

Sze SM. 1969. *Physics of semiconductor devices*. New York: Wiley International.

Van Duyneveldand DJ, Pouw CLM, Breur W. 1972. *Physica* **57**:205.

Van Santen JA, Van Duynevelt AJ, Carlin RL. 1975. *Physica* **79B**:91.

Van Vleck JH. 1940. *Phys Rev* **57**:426.

Van Vleck JH. 1941. *Phys Rev* **59**:724.

Vigouroux B. 1973. La mesure des temps de relaxation spin–réseau inférieures à 10^{-8} s: application au Gadolinium métallique. Thèse de 3ème cycle, Université Paul Sabatier, Toulouse.

Waller I. 1932. *Z Phys* **79**:370.

QUANTITATIVE MEASUREMENT OF MAGNETIC HYPERFINE PARAMETERS AND THE PHYSICAL ORGANIC CHEMISTRY OF SUPRAMOLECULAR SYSTEMS

Christopher J. Bender

Department of Chemistry, Fordham University, 441 East Fordham Road, Bronx, New York 10458, USA

1. INTRODUCTION

The physical principles that underlie organic reactions were established by a systematic study of chemical reaction dynamics that employed correlated measurements of reaction rates and a physical parameter that could be related of the electronic properties of the molecules in question (Hammett, 1970). Today, molecular science emphasizes the concept of molecular device, which connotes a supramolecular structure (the term "supramolecule" loosely means a molecule that has multiple functionalities associated with it; for example, an enzyme might be regarded as a supramolecule in the sense that it features a supported metal catalyst and a receptor site that recognizes a specific substrate upon which the catalyst acts) that acts in some specific fashion. A molecular device may be biological (*e.g.*, enzymes, contractile proteins; *cf.* Tanford & Reynolds, 2001), or it may be produced by synthetic means (*e.g.*, molecular wires, switches, machines, etc.; *cf.* Sauvage, 2001; Balzani *et al.*, 2003). Current synthetic chemistry provides the technical means that enable one to create and modify molecular devices so that structure may elicit some specific function, and so physical organic chemists are interested in reactions that involve engineered and structurally complex systems such as supported catalysts, protein active sites, or nanostructures (*cf.* Hamilton, 1996; Tidwell *et al.*, 1997).

Our understanding of small molecule reaction dynamics is made possible by combining the results of theoretical chemistry, spectroscopy, and mechanistic organic chemistry. This approach, which was successful when applied to small molecules, is, in principle, applicable to supramolecular entities, but there arises in this latter case the question as to whether there can be obtained enough accurate information from spectra in order to make meaningful correlations to chemical behavior. Magnetic resonance parameters, such as the chemical shift, nuclear hyperfine interaction, and nuclear quadrupole interaction, are useful probes of molecular electronic structure (*cf.* Memory, 1968; Semin *et al.*, 1975; Harriman, 1978; Ando, 1983; Kaupp *et al.*, 2004), and they have consistently been used in correlated structure–function studies and probes of molecular interactions (*cf.* Ratajczak & Orville-Thomas, 1982; Schuster *et al.*, 1976).

When applied to large molecular systems such as an enzyme or a supramolecular entity, however, methods such as vibrational spectroscopy and magnetic resonance often suffer from an overload of information (too many spectral lines) and/or a loss of information (*i.e.*, precision) due to line broadening. With large molecules, it is often not possible to obtain sample materials as single crystals or homogeneous solutions, and so one obtains spectra that represent an averaging over all possible molecular orientations (relative to a laboratory reference frame) and subject to certain long-range interactions, thus losing the ability to detect small changes in spectroscopic features. What is needed for these so-called supramolecular systems is a spectroscopic probe that is selective towards the chemically interesting region of the molecule (*i.e.*, self-editing) and has available a means to defeat the line-broadening effects that result in spectroscopic information loss. The premise behind this chapter is that two advanced methods of electron magnetic resonance, namely, electron-nuclear double resonance (ENDOR) and electron spin echo envelope modulation (ESEEM), are ideal spectroscopic tools for correlated studies of chemical behavior involving such supramolecular entities.

Many supramolecular entities, whether biological or synthetic, have some sort of paramagnetic species (transition metal, organic free radical) as its chemically reacting agent, the remainder of the structure serving as a kind of support or package. Electron magnetic resonance (EMR) is sensitive to changes in the electron spin state of paramagnets, and so in this sense it is self-editing the spectroscopic information obtained to the chemical agent of interest and its immediate environment (typically within 10 Å). The interaction between the unpaired electron and its local environment yields nuclear hyperfine interaction and nuclear quadrupole interaction that are recovered from perturbations to the EMR (electronic Zeeman) spectrum and are manifestations of the magnetic interaction between the atomic nucleus and the valence electron orbitals, establishing their relevance as probes of electronic structure. But in conventional EMR spectra of large molecules one does not typically obtain the requisite spectral resolution to measure small changes in the hyperfine parameters.

ENDOR and ESEEM are specialized EMR methods that, in effect, decouple the measurement of nuclear hyperfine interactions (broadly defined) from the electronic Zeeman transition. In other words, rather than viewing the paramagnet's environment as a perturbing influence on the electron spin "flip" (*i.e.*, a small perturbation on a much larger energy transition), one records the nuclear hyperfine spectrum of the environment indirectly as though it were an NMR experiment by using one of the electron spin states as an observer. So instead of so-called hyperfine lines in the oftentimes broadened spectral line corresponding to the Zeeman transition, one measures the nuclear hyperfine spectrum directly, and as a result, one gains over three orders of magnitude better resolution and linewidths on the order of 10 kHz. Since many of the parameters we wish to measure have values on the order of 1–10 MHz, the ENDOR and ESEEM methods enable one to resolve differences, either as peak shifts or separate peaks, on the order of 10 kHz. And this means that, in principle, we can determine nuclear quadrupole interaction parameters, nuclear hyperfine coupling (contact and magnetic dipole), etc. to a precision that is suitable to our stated application.

EMR spectra, including ENDOR and ESEEM, however, have the disadvantage of the fact that the microwave source used in the typical spectrometer operates at a single frequency, for example, approximately 9.5 GHz for an X-band spectrometer. This is a problem because the spin-Hamiltonian description of the EMR spectrum is a sum of several parameters that are tensors, and so that interpreting and assigning these parameters from a single EMR, ENDOR, or ESEEM spectrum involves both energetic and geometric (*i.e.*, crystallographic) parameters. These parameters are typically assigned in this case by comparing the experimental spectrum to simulated spectra obtained via a trial-and-error fitting of the spin-Hamiltonian parameters and their respective tensor orientation. And so with only a single experimental spectrum one is faced with more independent theoretical parameters (*i.e.*, spin-Hamiltonian terms and tensor angles) than there is information in the spectrum. The trial-and-error fitting of simulated to experimental spectra is subjective and detracts from the confidence one would have in assigning spectroscopic parameters in spite of the claims I have made in the previous paragraph concerning the resolution of the spectra themselves.

The situation is improved by recording spectra at multiple spectrometer operating bands (*e.g.,* C-, X-, P-, Q-, *etc.*) and simulating each of these spectra with the same set of spin-Hamiltonian parameters. But this type of multi-frequency approach leaves one with a series of experimental spectra that are often disparate in profile, making it difficult to discern trends. But it is possible to eliminate much of the ambiguity in assigning hyperfine terms by performing a multi-frequency recording of spectra in small incremental steps (*ca.* 500 MHz) near the region where the nuclear hyperfine states of one electron spin manifold merge, which is the condition of "exact cancellation" in the jargon of ESEEM (*cf.* Mims, 1972b,c; Singel, 1989). This technique is analogous to those found in other branches of spectroscopy, going by the name of, for example, quantum beats spectroscopy, level (anti-) crossing spectroscopy. The upshot is that the quantum beats phenomenon leads to extremely narrow lines, and that when these line positions are plotted against the nuclear Larmor frequency, a graphical procedure enables one to assign the energy terms of the nuclear hyperfine interaction more accurately (*cf.* Singel, 1989; Bender *et al.*, 1997; Bender & Peisach, 1998).

The purpose of the present chapter is twofold: first, to show that the EMR method is a powerful tool for the study of chemical reaction dynamics of large supramolecular entities in the same experimental style that was used by physical organic chemists working with small molecules. This will be outlined in §2. Second, I hope to successfully convince the reader that ENDOR and ESEEM spectra can provide sufficiently accurate and reliable information so that derived parameters such as nuclear hyperfine and quadrupole parameters can be applied in correlated studies of chemical dynamics involving supramolecular systems. This will be demonstrated by taking a different approach to the multi-frequency measurement of ENDOR and ESEEM spectra, that is, one in which multiple spectra are recorded at identical g-values but at applied magnetic field strengths that vary in small incremental steps, rather than single spectra in two or more bands; the spectral peak positions are then plotted as a function of the nuclear Zeeman energy. Doing so enables one to graphically identify the true condition of exact cancellation at which point zero-field nuclear quadrupole interaction (ZF–NQI) parameters may be assigned. The reason one needs this graphical approach (outlined in §5) is that dipolar broadening of the EMR line, together with the broad bandwidth of the excitation pulse (ESEEM), results in exact cancellation-like spectra over an extended range of nuclear Zeeman values, meaning that, unless one traces the movement of peaks as a function of nuclear Zeeman energy, it is impossible to determine the true point of exact cancellation.

The remainder of this chapter is organized as follows. Section 2 reviews the logic behind the so-called Hammett sigma analysis and describes prospective analogous experiments on metalloproteins that can be studied via ESEEM and ENDOR. Section 3 reviews the terms of the spin Hamiltonian and their connection to electronic structure theory. Section 4 compares the ENDOR and ESEEM techniques, examining their complementary qualities. Section 5 demonstrates how a graphical approach to the measurement of ESEEM or ENDOR peak position greatly facilitates the assignment of exact cancellation.

2. SPECTROSCOPY IN CORRELATED CHEMICAL DYNAMICS

In principle, knowledge of an atom or molecule's electronic structure (*i.e.*, the quantum mechanical wavefunction) would enable one to predict both its physical properties and its chemical behavior, including the outcome of reactions with other atoms or molecules whose electronic structure are equally well known (*cf.* Daudel, 1973; Daudel *et al.*, 1982). But because the Schrödinger equation cannot be solved exactly for any system more complicated than the hydrogen atom, the wavefunction of atoms and molecules must be approximated. Spectroscopy provides us with an observational link between the macroscopic and microscopic realms of matter, and it has been both a guide to our conceptual understanding of matter and a means to approximate parameters that are used in semiempirical computational chemistry (*cf.* Segal, 1977).

This same tactic was applied to the analysis of chemical reaction dynamics. Lacking a complete quantum mechanical description of molecules, chemists recognized that they could alter the electronic structure of a molecule by functionalizing it with nonparticipating substituents. The degree to which the substituents perturbed a molecule's electronic structure could be assessed via spectroscopy and made quantitative in the form of a so-called substituent parameter (*cf.* Hammett, 1970). The principal requirement of the chosen spectroscopic parameter is that it be rooted in the electronic structure of the molecule. The magnetic hyperfine terms that comprise the EMR spin Hamiltonian all qualify in this regard (Bowers, 1968; Viehe *et al.*, 1986; Jiang, 1997).

For small molecules, structure–reactivity analysis proceeds by synthesizing a series of substituted variants of the parent molecule (*e.g.*, substituted benzenes, *cf.* Brown & Goldman, 1962), which are then made to react in some identical manner. The information that one obtains consists of reaction rates for each of the molecular variants and some physical parameter, selected according to the criterion cited in the previous paragraph, measured for each variant. The interrelationship between chemical reaction rate and electronic structure is then parametrically written as

$$\log k_{ij} - \log k_{0j} = \sigma_i \rho_j \qquad (1)$$

where term σ_i is denoted as the substituent constant, which is correlated to some physical property of the molecule(s), and ρ_j is the reaction parameter. Subscripts i and j designate a specific substituent and type of reaction, respectively. Thus k_{0j} is the rate constant of reaction j when the parent molecule lacks a substituent, and k_{ij} is the reaction rate constant that is obtained when substituent i is attached to the parent molecule.

This same analytical procedure is now, in principle, applicable to supramolecular structures, including proteins. Many of the supramolecular devices of interest, natural and synthetic, are templated, and can therefore be modified in a systematic manner. For example, an enzyme is templated by a nucleic acid sequence and may be synthesized in large quantities *in vitro* by now routine molecular biology methods, such as the polymerase chain reaction (PCR). Site-directed mutagenesis enables one to selectively modify the nuclei acid sequence and therefore the protein in a controlled manner. And so it is possible (in fact, routine) to systematically modify the region local to the enzyme's active site by replacing one amino acid (*e.g.*, the wild type) by another (site-directed mutagenesis), and then observing changes in chemical behavior and physical properties.

Consider cupredoxins, which are small metalloproteins that shuttle redox equivalents among membrane bound proteins of biological electron transport chains. The question of redox tuning (specifically, what manner of molecular interactions within a protein (or its cofactors) determine its standard reduction potential) has been a long-term problem in bioenergetics, and so one might devise experiments in which the amino acid ligands to copper or amino acids near the active site are modified and then observe the effect of these modifications on the electron transfer self-exchange rate. An experiment of this type was performed in order to determine the path of electron transfer between the copper ion and the periphery of azurin (*cf.* van de Kamp, *et al.*, 1993), but in this precedent ESEEM was used only as a diagnostic tool to verify that the copper binding site had not been altered by an amino-acid substitution in the protein milieu.

But even more ambitious probes of the structure–function correlation are afforded by molecular biology. There exist techniques that enable one to introduce non-natural amino acids into a protein sequence (Cornish *et al.*, 1995). This means that the chemical reactions of enzymes may be studied in precisely the same manner as small molecule reactions were studied during the 1960s and 1970s. Using cupredoxin as a hypothetical example, the incorporation of non-natural amino acids makes it feasible to synthesize (for example) histidine with various substituents on the imidazole ring and use these to alter the copper binding site of a cupredoxin and systematically measure their impact on electron transfer. The significance of the non-natural amino acids is this: with site-directed mutagenesis one substitutes one amino acid for another, and thereby potentially changes the chemistry of the process; non-natural amino acids allow one to work with variants of the wild-type amino acid and merely perturb the chemical process.

EMR is an excellent tool for correlated structure–function experiments and the analysis of chemical reactions. It may be used for the measurement of reaction kinetics (*cf.* Levanon & Möbius, 1997; Clancy *et al.*, 1998; Grampp, 1998; Murai *et al.*, 2000), (crystallographic) structural analysis (Box, 1977; Weltner, 1983), and as a probe of electronic structure (Harriman, 1978; Mabbs & Collison, 1992; Deligiannakis, *et al.*, 2000). For many important chemical reactions, a paramagnetic metal ion, free radical, or organic molecule in its triplet state participates as a reactant or intermediate, and EMR is very often the spectroscopic tool of choice for the analysis of such reactions because the EMR spectrum is only sensitive to the paramagnet and a spatially limited number of magnetically coupled nuclei. In other words, EMR provides a spectroscopic "window" onto the chemical species of interest and its immediate environment, and the technique is therefore minimally affected by molecular size and sample heterogeneity.

The significance of this spectroscopic localization inherent to EMR may be appreciated by examining Figure 1, which depicts the ligand binding motif of a cupredoxin, namely azurin. The Cu(II) ion is the chemically interesting (paramagnetic) entity whose EMR spectrum contains so-called "hyperfine structure" that is indicative of magnetic interactions between the unpaired electron (on copper) and nitrogen atoms in the coordination sphere. This hyperfine interaction is defined as a series of terms (*i.e.,* the spin Hamiltonian), and one of these terms is the nuclear quadrupole interaction (NQI), which is regarded as one of the most useful probes of local electronic structure (Lucken, 1969a,b; Semin, 1975; Guibé & Jugie, 1981; Gordy & Cook, 1984). Looking at the representative Type I copper site, we see that, besides copper itself, there are several potential NQI probes of the metal site (^{14}N, ^{33}S). The table that accompanies Figure 1 catalogs the common amino-acid ligands to metals in metalloproteins (Adman, 1991), and again one notes that quadrupolar nuclei are prevalent. This means that EMR spectroscopic recovery of NQI information is generally applicable in metalloproteins and that this information may, in principle, be used for correlated analyses of enzymatic (or, in general, supramolecular) reaction dynamics.

One needs, however, to ensure that the spectroscopic probe of our system provides information (in the form of spectroscopic peaks) that is energetically commensurate to the perturbations one expects to see as a result of inter- or intramolecular interactions. And here it is worthwhile to examine the energetics of

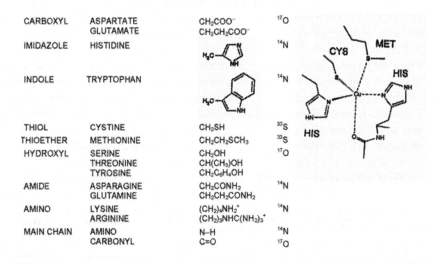

Figure 1. Amino acid ligands to metal ions in proteins, and the quadrupolar nuclei that may be used as local probes of electronic structure (after Adman, 1991). On the right, a Type I copper site (azurin) as a representative example (after Baker, 1988).

molecular interactions in general for a comparative sense of the various spectroscopies that one may use. First, one needs to define the interaction energy itself. Using *ab initio* quantum mechanics as a guide, one recognizes two states: one corresponding to the isolated atoms (or molecules), and a second describing the complex of interacting atoms. The interaction energy is then defined as the difference between these two state energies, and this is computationally problematic because the small interaction energy is being defined as the difference between two large numbers and so leads to substantial error (*cf.* Boys & Bernardi, 1970).

The difficulty of this definition (or detection) of interactions is one of scale; the perturbations that one is trying to observe are orders of magnitude smaller than actual measurements. Computationally, a better approach is to attempt evaluation of the individual perturbations directly, and then define the total interaction energy as a sum of the individual perturbation energies. In the case of EMR spectroscopy, this is exactly what we are doing by using ENDOR or ESEEM. We know that the effects that we expect to see will become manifest in the nuclear hyperfine terms, so rather than try to measure these from differences in the EMR spectrum, which includes the electron Zeeman term, we turn instead to the ENDOR and ESEEM, which detect the nuclear hyperfine interactions.

Regarding a matching of energetic scale, we can look to representative data from the literature of molecular interactions. For example, a variational decomposition scheme for the *ab initio* calculation of specific interaction energies can be used as a starting point for determining the magnitude of interaction energies. In the decomposition scheme of Morokuma and Kitaura (Morokuma, 1971; Kitaura & Morokuma, 1976), the (molecular) interaction energy may be written as a sum of electrostatic (ΔWES), exchange (ΔWEX), charge-transfer (ΔWCT), polarization (ΔWPL), and mixing (ΔWMIX), with each of these terms corresponding to a unique integral expression (the Slater determinant). Table 1 lists computational data obtained by Morokuma and Kitaura (1980) comparing the changes in the intermolecular interaction energy (*i.e.*, $\Delta\Delta$W *vs.* ΔW) for several complexes following a replacement of H– by a CH_3–, in other words, the change of an interaction energy as a substituent effect. The energies are reported in units of cm^{-1} (converted from kcal·mole^{-1}) in order to facilitate comparisons to spectroscopic data.

Table 1. Changes in the Variational *ab initio* Interaction Energy Resulting from a Methyl Substitution (adapted with permission from Morokuma & Kitaura, 1980)

	$H_3N–OH_2$	$H_3N–ClF$	$H_3N–BH_3$	$H_3N–Li^+$	$H_3N–H^+$
R_e (Å)	2.93	2.717	1.705	2.01	1.705
ΔW (cm^{-1})	–3150	–2870	–15644	–17779	–77662
Methyl substituent effect					
$\Delta\Delta$W	70	105	–280	700	–2975
% Change	+2%	+4%	–2%	+4%	–4%
$\Delta\Delta W_{ES}$	105	105	–420	1470	1155
$\Delta\Delta W_{EX}$	70	175	1540	140	0
$\Delta\Delta W_{PL}$	–35	0	–1750	–910	–4480
$\Delta\Delta W_{CT}$	–70	–210	–490	–70	–1190
$\Delta\Delta W_{MIX}$	–35	0	840	70	1540

The molecular interaction energies are on the order of 10^3–10^4 cm^{-1}, which corresponds well to peaks such as those seen in the vibrational spectroscopy of, for example, hydrogen-bonding complexes (*cf.* Bratos *et al.*, 1980; Janoschek, 1976). The changes that accompany the replacement of H– by a CH_3–, however, are one or two orders of magnitude smaller. In solids, the infrared and Raman half-widths of lines corresponding to weak hydrogen bonding interactions are typically 20 cm^{-1} (Janoschek, 1976), and so one is at perhaps an upper limit spectroscopically with respect to detecting weak interactions. As one moves to higher frequencies (submillimeter, microwaves, radio waves), the perturbations become significantly larger than the energies of the spectroscopic transitions and differences more easily observed.

Numerous spectroscopic probes are used to assess intra- and intermolecular interactions (*cf.* Ratajcak & Orville-Thomas, 1980), of which the magnetic resonance methods give highest resolution. An early demonstration of the NMR-detected substituent effect was published by Gutowsky *et al.* (1952), which yielded shifts of the ^{19}F resonance lines ranging from -16 to $+10$ ppm subject to meta- and para-substituents on fluorobenzene. Webb and Witanowski (1985) have extensively catalogued the molecular interaction effects on ^{14}N shielding via NMR, with shifts often measured at 10–100 ppm with a precision of 0.1 ppm (in solution samples). Likewise, Guibé and Jugie (1980) catalog nuclear quadrupole spectral studies of molecular complexes; these typically measured on the 1–100 megahertz scale, with peak shifts due to interaction typically being about 10%.

Since we are hoping to use quadrupolar nuclei near a paramagnetic chemical entity of interest, it follows from the preceding paragraphs that, on one hand, EMR spectroscopy, in general, is well matched energetically to weak forces between molecules. Second, the information (*i.e.,* hyperfine parameters) obtained from advanced EMR methods such as ENDOR and ESEEM permit one to record nuclear hyperfine spectra (analogous to NMR or NQR) whose parameters are traceable via the spin Hamiltonian to electronic structure (§3). Furthermore, under optimal conditions (coherence transfer, *g*-selection, *etc.*), both ENDOR and ESEEM yield powder-pattern spectra whose lines are resolvable to the tens of kilohertz ($\sim 10^{-7}$ cm^{-1}) and therefore sufficiently precise to detect small shifts in the nuclear quadrupole interaction. This matching of the spectroscopic energetics to that of the magnitude of the perturbations one expects to impose is important in order to make a quantitative evaluation of the substituent effect(s).

3. THE SPIN HAMILTONIAN

3.1. Perturbational Expansion and Decomposition of Spin Interaction Energies

Electron magnetic resonance is a form of microwave spectroscopy, but electron magnetic resonance spectra correspond to transitions among states associated with a change in the molecular spin magnetic moment. The correlation of spectroscopic data to chemical properties is best described in terms of a spin-Hamiltonian model that is derived from the magnetic properties of the electron. This spin Hamiltonian is written as a perturbational expansion, and each of its terms may be correlated to a chemically relevant interaction of the electronic valence shell. The deconvolution of individual spin-Hamiltonian terms may be complicated and prone to systematic error, but the magnetic field dependence of at least one contributing term can be used as a powerful tool in the sense that the spectrum can, in principle, be advantageously "tuned" so as to bring into play phenomena that greatly simplify its interpretation (§5).

At moderate to high magnetic fields (*i.e.,* $H_0 \sim 0.1$ T), the electron Zeeman interaction dominates the EMR spectrum, which at its simplest may be described by a single transition $h\nu = g_e \beta_e H_0$, where g_e and β_e are the so-called g-value and the Bohr magneton (A fundamental unit of the electron's magnetic moment, $e\hbar/2m_e = 9.274 \times 10^{-24}$ J·T^{-1}), respectively. In atoms and molecules the g-value is replaced by a tensor and deviates from the scalar quantity of 2.0023 for a free electron. The deviation between the observed g and g_e reflects the spin–orbit interaction of the electron because the electron's orbital motion produces a magnetic field opposing the applied field; detailed studies of these g-value variations are now conducted by using high (Zeeman) field EMR spectroscopy (*cf.* Budil *et al.*, 1989; Un *et al.*, 1994; Prisner, 1997).

Simple $g_e \beta_e H_0$ Zeeman splitting is an inadequate model of the observed electron magnetic resonance spectrum in molecular systems because the unpaired electron spin interacts with its local magnetic and electrical environment, and the magnitude of this interaction energy is significant on the scale of the electron Zeeman energy. Fine structure therefore appears in the microwave Zeeman spectrum, and the spin Hamiltonian is expanded as a sum of interaction energies. Common interactions include those between the magnetic moments of the electron and nearby atomic nuclei (or other unpaired electrons), but internal magnetic and electric fields will likewise affect the fine structure spectrum. The expanded spin-Hamiltonian terms are best understood as the quantum mechanical or semiclassical analogues of electromagnetic interactions, which are expressed classically as vector product sums. The electron and nuclear spin magnetic moments are $g_e \beta_e S$ and $g_n \beta_n I$, respectively, where S and I are the electron and nuclear angular momentum, and the total energy is decomposed into several conceptually convenient terms, as is done classically (*cf.* Becker, 1964). For an electron coupled to one or more nuclei the spin wavefunction is written as a product of $|m_S\rangle$ and all (coupled) $|m_I\rangle$, and the two terms of the spin Hamiltonian account for the electron and nuclear Zeeman interactions. Additional terms are added as needed in order to describe electromagnetic interactions between the electron and nuclei in a descending order of their magnitude, and the general form (Poole & Farach, 1987) of the spin Hamiltonian is therefore

$$\mathbf{H}_{spin} = \beta_e \mathbf{S} \cdot \mathbf{g}_e \cdot \mathbf{S} + \beta_n \mathbf{I} \cdot \mathbf{g}_n \cdot \mathbf{I} + \mathbf{S} \cdot \mathbf{A} \cdot \mathbf{I} + \mathbf{I} \cdot \mathbf{P} \cdot \mathbf{I} \qquad (2)$$

where each term represents the product of second-order (*i.e.*, rank) tensors $\mathbf{g}_e$, $\mathbf{A}$, and $\mathbf{P}$, and Pauli spin matrices $\mathbf{S}$ and $\mathbf{I}$. Figure 2 graphically depicts the hyperfine energy decomposition as per the spin-Hamiltonian formalism for the interaction between a single unpaired electron, $S = 1/2$, and a nucleus of spin $I = 1$, such as ^{14}N. The spin states are designated by using the ket convention $|m_S, m_I\rangle$. From left to right in eq. (2), the terms are defined as the electron Zeeman, nuclear Zeeman, nuclear hyperfine (a sum of an isotropic and purely quantum mechanical contact interaction and the anisotropic magnetic dipolar interactions), and nuclear quadrupole coupling interactions. Other terms may be added as appertains the spin system of study (*cf.* Poole & Farach, 1987), but the weak hyperfine terms involving electron–nuclear interactions are the subject of this review and will be described fully in turn.

The energy splittings of Figure 2 are not drawn to scale. The electron Zeeman splitting in a $0.1 - T$ magnetic field is 2.8 GHz (~0.1 cm^{-1}). The nuclear Zeeman, Fermi contact, Quadrupole, and dipole interaction energies are several orders of magnitude smaller and measured on the megahertz energy scale. Conventional

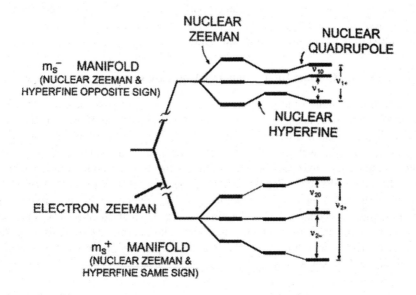

Figure 2. Representative energy level diagram for an S=1/2, I=1 spin system, subject to the spin Hamiltonian. Transitions indicated correspond to those observed in a nuclear hyperfine spectrum, such as that obtained via ESEEM.

EMR spectra correspond to transitions among the electronic Zeeman levels, subject to selection rule $\Delta m_S = \pm 1$, $\Delta m_I = 0$ (for all m_I of the system), and it follows that these comparatively large transition energies are difficult to correlate with weak molecular interactions or subtle changes in the nuclear hyperfine energies due to substituent effects.

By contrast, however, the advanced EMR techniques ENDOR and ESEEM are subject to different selection rules, that is, $\Delta m_S = 0$, $\Delta m_I = \pm 1$, and so ENDOR and ESEEM spectra correspond to transitions among nuclear sublevels (indicated in Figure 2). These transitions occur on an energy scale that is orders of magnitude smaller than conventional (*i.e.*, electronic Zeeman) EMR transitions and commensurate to the energies of the weak perturbations we desire to observe. The inherently narrower lines and finer energy scale of ENDOR or ESEEM spectra better enable one to detect small shifts of the peak position that accompany weak perturbations to electronic energy such as those that may accompany a substituent modification on a free radical or metal ion ligand.

In effect, ENDOR and ESEEM spectra permit one to conceptually drop the electronic Zeeman term from the spin Hamiltonian and work on an energetic scale that is comparable to NMR spectroscopy. Both techniques can be used to obtain specific terms of the spin Hamiltonian provided that one has the means to experimentally deconvolute the spectroscopic transition energies. The primary difference between ENDOR and ESEEM resides with the manner in which the electronic

Zeeman contribution is effectively taken out of the spectrum. In the remainder of this section, the chemically relevant spin-Hamiltonian terms will be described in the context of their chemically relevant origin.

3.2. The Fermi Contact Interaction and Magnetic Resonance Parameters

The fine structure of atomic line spectra and the hyperfine spectra of the electronic Zeeman effect is characterized by a symmetric splitting of the primary lines. The nuclear Zeeman interaction symmetrically splits the spectral line, and it is easy to predict from the well-established nuclear Zeeman relationship, $g_n\beta_n H_0$, which kind of nuclei are involved, but deviations of the fine structure splitting pattern from the expected nuclear Zeeman splitting suggested that there exists other interactions between the electron and atomic nucleus. This correction is derived from purely quantum mechanical considerations and is a scalar quantity that is defined as a measure of an atomic nucleus magnetic field (Fermi, 1930). This so-called contact energy is measured from the hyperfine splitting of the electron Zeeman transitions and is expressed as:

$$a = -\tfrac{8\pi}{3} g_e g_n \beta_e \beta_n \left| \Psi^2(r=0) \right| \tag{3}$$

where $|\Psi^2(r=0)|$ is defined as the probability of finding the unpaired electron (often called the electron density) at the atomic nuclear center. This interaction energy may, in principle, be numerically computed via quantum mechanics by representing Ψ as a radial wavefunction, and because Ψ is a radial wavefunction, the contact energy should vanish for situations in which the unpaired electron occupies an orbital of nonzero angular momentum (i.e., orbitals that are not s or σ). The contact interaction is therefore a measure of the s-character in the valence shell of the paramagnetic species.

The term $|\Psi^2(r=0)|$ in the definition of contact interaction is at odds with many observed EMR spectra of free radicals whose ground-state electronic configuration would necessarily put the unpaired electron unpaired electron in a p- or π-orbital. A similar argument holds for the d-orbitals of transition ions, whose EMR spectra are likewise interpreted by the occurrence of a contact interaction. The requisite s-character is introduced via an admixture of excited-state orbitals that is known as configuration interaction (Weissman, 1956; McConnell, 1956; Jarrett, 1956). Although the concept is now commonplace in *ab initio* quantum chemical technique, configuration interaction was introduced to account for the contact hyperfine interaction that was observed in the EMR spectra of π-radicals and paramagnetic ions. It follows, therefore, that the contact interaction can be used as one experimental measure of wavefunction accuracy in the specification of density matrices via configuration interaction during *ab initio* quantum chemical procedures.

In order to compute the contact interaction, the Slater determinants of the ground and possible relevant excited states (i.e., sets $\phi_i \mid i \geq 1$ are written and combined as a symmetry-adapted sum:

$$\Phi = \phi_0 + \lambda_1\phi_1 + \lambda_2\phi_2 + \lambda_3\phi_3 + \cdots \tag{4}$$

Coefficients λ_i are determined by applying perturbation theory, with the result that the ϕ_i terms are written as the product of an exchange integral (between occupied and virtual orbitals) and a quantity denoted as the spin density localized at the atomic center. The corresponding perturbational energy, the contact energy, is therefore expressed in terms of this spin density, that is, $a = Q\rho$, where ρ represents the spin density. The integral terms of the solution to the perturbational solution for energy and other coefficients are lumped as a single constant, Q, that is typically used to describe a given class of contact interactions and therefore empirically determined for such a class (*cf.* Gordy, 1980).

3.3. The Magnetic Dipole Interaction

The contact interaction energy is ordinarily combined with a tensoral energy that represents the dipole–dipole interaction between the magnetic moments of the unpaired electron ($\mu_S = g_e\beta_e$) and the (local) atomic nuclei ($\mu_I = g_n\beta_n$). The scalar expression for the dipolar interaction energy is

$$A_\mu = g_e g_n \beta_e \beta_n \left(\frac{1-\cos^2\phi}{r^3}\right) I \cdot S \tag{5}$$

where ϕ_i represents the angle that is formed between the vector connecting the nucleus and unpaired electron and the vector that defines the direction of the applied magnetic field. This angle introduces an orientation dependence to the spin-Hamiltonian term and permits one to assign spatial parameters r and ρ_i, the latter being the spin density assigned to atomic center i.

A very useful application of the dipole interaction tensor is the use of the electron–proton interaction to refine the spin density distribution in a delocalized system and, in turn, accurately determining McConnell constant Q. For example, given an isotropic contact interaction a, one may only estimate ρ based on an average Q that is derived from a matched class. The dipolar tensor for a given hydrogen atomic center, however, is the sum of all dipolar interactions over extended spatial coordinates of the coupled nuclei. The individual dipolar coupling energies, $A_{\mu,i}$, may be computed and summed into a common tensor reference frame. A fitting procedure may therefore be used to refine the spin density distribution throughout the atomic nuclear centers, as first described by McConnell and Strathdee (1959) for aromatic systems. The procedure has been used with dipolar couplings culled from ENDOR powder spectra to assign spin density distributions in the semiquinone anion radical (O'Malley & Babcock, 1980) and the tyrosyl radical of ribonucleotide reductase (Bender *et al.*, 1989). In the latter study, the neutron diffraction structure of tyrosine was used to define the geometry over which the electron–nuclear interactions were mathematically modeled, but an excellent fit to the experimental data within the error limits of the experiment could be obtained by using a geometry predicted from standard bond lengths and angles. It follows that the principles of McConnell-Strathdee analysis are generally applicable even in cases when the precise geometry of the delocalized spin system is not known.

The dipolar interaction of protons with nearby unpaired electron spin density is also useful as an aid to estimating orbital hybridization on carbon and nitrogen atomic centers. The McConnell Q tends to work well for interactions of a specific class, which among protons are the familiar α-, β-, and γ-proton classes, but Q may or may not be a single-valued scalar quantity. For example, hyperconjugation effects are manifest in the Q of β-protons as a sum: $Q = B_0 + B_1 \cos \varphi$, where φ is the dihedral angle made between the C_β–H bond and the $C_\alpha p_z$-orbital (Gordy, 1980). Orbital hybridization in the valence shell of carbon and nitrogen likewise affects the orbital character, and Q might be expressed as a sum of weighted terms Q_i that correspond to various orbital types on a given atomic center (cf. Gordy, 1980). Under these circumstances, estimates of spin density (and perhaps Q) may be improved by examining the proton–dipole interactions, which provide a measure of the total spin density on the nearby center independent of quantum mechanical parameters (i.e., Q).

3.4. The Nuclear Quadrupole Interaction

The fine structure of atomic line spectra and the hyperfine splittings of electronic Zeeman spectra are non-symmetric for those atomic nuclei whose spin equals or exceeds unity, $I \geq 1$. The terms of the spin Hamiltonian so far mentioned, that is, the nuclear Zeeman, contact interaction, and the electron–nuclear dipolar interaction, each symmetrically displace the energy, and the observed deviation from symmetry therefore suggests that another form of interaction between the atomic nucleus and electrons is extant. Like the electronic orbitals, nuclei assume states that are defined by the total angular momentum of the nucleons, and the nuclear orbitals may deviate from spherical symmetry. Such non-symmetric nuclei possess a quadrupole moment that is influenced by the motion of the surrounding electronic charge distribution and is manifest in the hyperfine spectrum (Kopfermann, 1958).

The chemical relevance of the nuclear quadrupole interaction (NQI) becomes evident from a description of its phenomenological origin. The non-spherical nuclear charge distribution is coupled to the electronic charge distribution, and if the latter is likewise non-spherical, the interaction energy is subject to change if the nuclear (or electron) spin orientation is flipped. This non-spherical charge distribution is therefore a requisite condition for the nuclear quadrupole interaction (Evans, 1955), and is the underlying reason that NQI is an aid in probing the electronic orbitals of molecules and the effect of molecular interactions (Meal, 1952; Guibé & Juglie, 1981; Weiss & Wigand, 1990). The requisite that both the nuclear and the electronic charge distributions deviate from spherical symmetry limits quadrupole coupling to interactions between the nucleus and the valence electrons, because closed shell atoms feature a spherical charge symmetry. This implies that s-orbital electrons do not contribute to a nuclear quadrupole coupling, and d- and f-orbitals are too far extended to yield a nuclear quadrupole interaction (Wilson, 1952). In the context of spin-Hamiltonian formalism, we can recognize that the contact (s or σ character only) and nuclear quadrupole interaction (non-spherical valence orbitals), in principle, provide complementary information and a useful spectroscopic tool for examination of orbital interactions in chemistry.

The tensor component of the nuclear quadrupole interaction corresponds to an electric field gradient (EFG) that reflects the valence shell configuration of the atomic nucleus. The classical interaction energy is (Davies, 1967)

$$W = \sum \left[e_i \phi_i - \mu_\alpha^i E_\alpha^i + \frac{1}{3} \Theta_{\alpha\beta}^i q_{\alpha\beta}^i \right] \qquad (6)$$

which is typically simplified for nuclei in an axial orbital system as

$$\mathbf{H} = \frac{e^2 Qq}{4I(2I-1)} \left[3I_z^2 - I^2 + \frac{\eta}{2}(I_+^2 + I_-^2) \right] \qquad (7)$$

The resultant zero-field energy levels of ^{14}N ($I=1$) are (cf. Lucken, 1969b)

$$W_\pm = \tfrac{1}{4} e^2 Qq[1 \pm \eta] \qquad (8a)$$

and

$$W_0 = \tfrac{1}{2} \eta e^2 Qq \qquad (8b)$$

From the experimental standpoint, a system such as ^{14}N is well parameterized because there will, in principle, be three spectroscopic transitions from which it will be possible to determine the two NQI parameters, $e^2 Qq$ and η. Other spin systems, such as ^{17}O ($I = 5/2$), ^{33}S, and ^{35}Cl ($I = 3/2$) are more difficult to parameterize because there are fewer spectroscopic transitions than theoretical parameters, but techniques such as double resonance enable one to get the requisite information (Semin et al., 1975; Lucken, 1969b). Simple orbital descriptions of the nuclear quadrupole coupling parameters and orbital hybridization have been reported (Cotton & Harris, 1966), and measures of the nuclear quadrupole interaction parameters and their tensor orientation can therefore be applied to solving problems of molecular structure and reactivity.

In brief, the reason that NQI parameters are so useful is because the tenets of molecular structure and reactivity that are used to describe chemistry are based on models that describe an arrangement of atoms subject to "character" in the sense that some constituents are charged, neutral, or in some way polarized. For example, chlorine-containing molecules are useful subjects for testing the correlation between nuclear quadrupole interaction parameters and bond properties because both extremes of bond character are easily measured; a purely ionic character assigned to chlorine would yield a nearly complete shell associated with chlorine and no nuclear quadrupole coupling, whereas a covalent bond, such as that of Cl_2, would represent another extreme value because of electron sharing. It follows therefore that a direct comparison of the chlorine NQI parameters from a molecule such as methylene chloride with the NQI parameters of molecular Cl_2 and NaCl would supply a relative measure of ionic and covalent character in the methylene chloride (Wilson, 1952). When applied in general terms, NQI parameters may be correlated to valence bond or molecular orbital descriptions of bonds (*cf.* Lucken, 1969b; Semin *et al.*, 1975). Given the manner in which NQI parameters are rationalized in terms of orbital interactions, it seems that modern computational methods of valence bond theory (*cf.* Gerratt *et al.*, 1997) are ideally suited to correlation of spectroscopic data and a quantum mechanical description of bond order parameters.

3.5. Non-Hamiltonian Factors that Influence EMR Spectra

3.5.1. Line Broadening & Selective Excitation

The ideal sample for obtaining solid-state EMR spectra and measuring small perturbations due to chemical effects would be a single crystal of well-defined symmetry and orientation with respect to uniform H_0 and H_1 fields. Such a sample would yield inherently narrow lines whose width depended only upon dipole–dipole interactions between like spins, spin–lattice relaxation, and molecular motion; so-called homogeneously broadened lines (Geschwind, 1967). For most practical situations, however, the sample material is either a magnetically dilute solid that cannot be crystallized or a solution that must be frozen and kept at low temperature during the EMR experiment because the electron spin relaxation times must be slowed. These samples do not possess any orientation preference, and the resultant EMR spectrum is inhomogeneously broadened because of differences among the internal effective fields at various sites. Information in the hyperfine spectrum is often lost from the cw-EMR powder pattern line because of this inhomogeneous broadening, and one of the earliest recognized features of ENDOR was a resolution enhancement that corresponded approximately to the ratio of the inhomogeneous line to the individual hyperfine lines (Hyde, 1967).

The resonance condition $h\nu = g\beta_e H_0$ is oversimplified in its scalar form because, unless the atom or molecule bearing the unpaired electron has spherical symmetry, g_e and H_0 are replaced by tensor **g** and 1×3 matrix **H**, respectively. If the components of tensor **g** are not aligned with H_z, then the effective field is represented by projection **g·H** or $|g| |H| \cos\theta$, where θ is the angle made between the two vectors. It follows that in a sample comprised of randomly oriented paramagnets the resonance line will be broadened because of the distribution of angles θ. Each infinitesimal fraction of molecules oriented with angle $\delta\theta$ can be imagined as an individual homogenously broadened spin packet, and the complete broadened spectrum represents the superposition of the collection of homogeneously broadened spin packets, that is, $L = f(\theta)\delta\theta$. The causes of inhomogeneous broadening are hyperfine interaction, anisotropy broadening, dipolar interactions between unlike spins, and magnetic field inhomogeneities.

The g- and superhyperfine tensors, in general, are characterized on the basis of symmetry, and their relative orientation is fixed in the molecular reference frame (usually defined by molecular symmetry axes). The Hamiltonian that describes the electronic Zeeman interaction in a principal axis frame (no terms of the form g_{ij}, where $i \neq j$) is

$$\mathbf{H} = \beta H \left(g_x S_x \cos\theta_x + g_y S_y \cos\theta_y + g_z S_z \cos\theta_z \right) \tag{9}$$

where term θ_i corresponds to the angle formed between the applied field and the g_i-axis. It follows that selection of a specific g-value (in a homogeneous line or single spin packet) could be achieved by an appropriate orientation between H and the g-axes. In inhomogeneously broadened systems the spin Hamiltonian would be expressed as a sum over all the spin packets, which, when integrated over all possible orientations and angles theta, yields an average that may be statistically weighted with respect to spin packet distribution by varying H. In other words, one obtains a powder pattern cw-EMR spectrum whose shape features "turning points" that are indicative of high populations of each g-value orientations (*i.e.,* population selective, see Mabbs & Collison, 1992). These turning points can be used to select and resolve hyperfine parameters and ESEEM spectra via the population selection method as outlined above.

The classic study of g-orientation selection of hyperfine tensor parameters is that described by Rist and Hyde (1970) for the axially symmetric copper complexes. Selection of nuclear hyperfine and quadrupole tensor terms in powder patterns is effected by the same means as described in the preceding paragraph. First, a high proportion of similar spin packets are "selected" by performing the ENDOR experiment at a region near the "wings" of the cw-EMR spectrum, that is, at either the low or high field extremum where only a few θ values contribute. With the spin packets selected in this manner one has, in effect, reduced the number of transitions being pumped (in the ENDOR or ESEEM experiment), and one is now left with the angular dependence of the hyperfine terms of the spin Hamiltonian. For example, in the preceding paragraph product $\mathbf{S \cdot g \cdot H}$ was written in terms of a principal axis and the angles between them; hyperfine term $\mathbf{S \cdot A \cdot I}$ is handled in a similar fashion, and in a g-selected ENDOR experiment one obtains single crystal-like hyperfine spectra because one is reducing the number of contributing orientations. The angular dependence among the tensors, from which g-selectivity arises, is derived from an analysis of single-crystal EMR spectra (Rist *et al.*, 1968). General procedures for analyzing powder pattern hyperfine spectra are based upon integration over angular representations of all possible tensor orientations within a symmetry framework (Dalton & Kwiram, 1972; Hoffman *et al.*, 1984; Hurst *et al.*, 1985).

The g-selection effect is also observed in the case of free radicals for which there is g-anisotropy. For example, the [1]H-ENDOR hyperfine spectra of the tyrosyl radical of ribonucleotide reductase exhibits dramatic selectivity by the g-selection technique. Figure 3 depicts the selectivity obtained near the so-called matrix region of the ENDOR spectrum. At $g = 1.99$ the matrix region, which corresponds to protons that are weakly coupled via dipolar interaction to the unpaired electron (as defined by Hyde *et al.*, 1968), achieve very narrow, almost single-crystal quality, linewidths from which numerical values are easily recovered. Likewise, the axial line corresponding to the 2,6 α-protons (*cf.* Bender *et al.*, 1989) is well-resolved at $g = 1.99$ and demonstrates profound g-selection (the 3,5 α-protons also demonstrate g-selection and also seem to be resolvable). These data clearly show that single-crystal-like narrow lines can be achieved in powder pattern ENDOR spectra despite a small g-anisotropy (and therefore one is not limited to large anisotropy cases such as Cu(II), as demonstrated in the Rist & Hyde 1970 study), and these lines can be "selected"; the key to such resolution is having the spectrometer bandwidth close to the width of a spin packet.

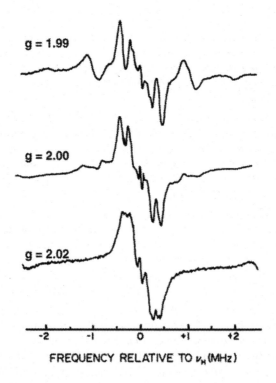

Figure 3. A proton (matrix) ENDOR spectrum of the tyrosyl radical of ribinucleotide reductase, illustrating both the high resolution of the technique and the ability to selectively enhance features in the spectrum by g-value (Bender *et al.*, 1989).

3.5.2. Spin Relaxation

Transitions among the various spin states are governed by rate equations that represent two counteracting probability functions. The first of these functions pertains to the radiation-induced transition and the probability that the time-dependent electric field of the incident radiation couples to the sample medium and induces transitions among the allowed spin states. The counterpart, spin relaxation, denotes the multifaceted dynamics of the spin population as it returns to thermal (*i.e.*, Boltzmann) equilibrium. The relative rates of these two processes determine the EMR signal intensity, and the associated spin dynamic processes serve as the basis for advanced EMR techniques and their specialized niches for determining super-hyperfine parameters.

Spin–lattice relaxation denotes all direct relaxation processes subject to the selection rules applied to EMR spectra, but more than one mechanism of spin relaxation is extant. It was experimentally established, for example, that the rate of return to a Boltzmann distribution following excitation depends upon the spacing of the spin system's energy levels, and that the spin population distributed among equally spaced energy levels returned to its Boltzmann equilibrium distribution faster than the corresponding unequally spaced system.

The more rapid relaxation rate, denoted as T_2^{-1}, is related to the concept of spin temperature and spectral diffusion. In the case of an equally spaced three-level system the rapid return to the equilibrium state occurs because the routes of energy "disposal" are identical, and mutual flips induced by dipolar interaction among the members of the population are isoenergetic (Bloembergen *et al.*, 1959). On the other hand, when energy levels are unequally spaced the spin system first comes into equilibrium with the lattice via electron–phonon interactions, and this slower spin–lattice relaxation rate is designated as T_1^{-1}. But the measurements of spin relaxation dynamics as a function of field (Bloembergen *et al.*, 1959) that led to the identification of these two relaxation processes also revealed an intermediate condition that occurs as unequally spaced levels are manipulated so that they gradually approach the condition of equality. In such a situation there occurs an intermediate cross-relaxation rate T_{12}^{-1}, such that $T_2^{-1} > T_{12}^{-1} > T_1^{-1}$ (Bloembergen *et al.*, 1959; Grant, 1964a–d). As will become apparent in the sections to follow, spin relaxation phenomena are important in governing the ability to observe the ENDOR effect and ESEEM.

4. ADVANCED EMR: ENDOR VS. ESEEM

In the preceding section, spin Hamiltonian terms corresponding to weak hyperfine interactions and their chemical significance were described. Conventional cw-EMR spectroscopy measures these interactions as a small splitting of the lines corresponding to transitions among the m_S states (*i.e.*, the electron Zeeman transitions), and these splittings are prone to being lost within the inhomogeneous line broadening of spectra derived from the random orientation of individual molecular tensors with respect to the applied field (§3.5). The selection rules of cw-ENDOR spectra, that is, $\Delta m_S = 0$, $\Delta m_I = \pm 1$, obviate the electron Zeeman transitions and, in effect, directly measure the hyperfine interaction. ESEEM spectra appear as though subject to the same selection rules, but the two techniques differ profoundly with respect to the manner in which the spin dynamics of the spin system are manipulated and the response detected. These differences, however, benefit the spectroscopist in regard to the type of information that is recovered.

4.1. cw-ENDOR

cw-ENDOR spectroscopy relies on a dynamic competition in populating the electronic states during continuous irradiation, but it is otherwise familiar in the manner of its execution because it is a swept-frequency experiment. One of the unique features of magnetic resonance spectroscopy is that the energy difference between the ground and excited states is so small that it is possible to easily saturate the system and effectively burn a hole in the spectrum at moderate powers of electromagnetic radiation. In other words, the dynamical competition between the radiation-induced transition probability and the non-radiative relaxation rates can be very easily controlled and therefore balanced. ENDOR is a variant of the Overhauser spectroscopic technique in nuclear magnetic resonance because spectra are derived from a manipulation of the electron and nuclear spin dynamics and their interaction (Dwek et al., 1969).

In order to prime the system for cw-ENDOR detection, an allowed EMR transition ($\Delta m_S = \pm 1$, $\Delta m_I = 0$) is saturated in the sense that ground- and excited-state populations are equalized, which means that the signal detected by the EMR spectrometer is rendered transparent (a "hole is burned" in the EMR spectrum). Irradiation of the sample with a second frequency that corresponds to the difference between nuclear substates will open up a new pathway for the spin transitions and depopulate one of the energy states participating in the saturated system. The radiation-induced depopulation of one state breaks the impasse between the EMR transition and the spin–lattice interaction rate. As the NMR transition depopulates the saturated states, the EMR signal is recovered and one records an NMR spectrum that reflects the hyperfine splittings of the nuclear sublevels (selection rule: $\Delta m_S = 0$, $\Delta m_I = \pm 1$). A schematic description of the routes of spin-state transfer is often drawn in analogy to a four-terminal electrical circuit and may be found in most reviews on the subject (Dwek et al., 1969; Kevan & Kispert, 1976; Schweiger, 1982).

The principal advantage of ENDOR spectroscopy is the much finer energy scale upon which the state-to-state transitions are recorded. As discussed in §3, conventional cw-EMR spectroscopy detects the weak hyperfine interactions of the spin Hamiltonian as a perturbation of the electronic Zeeman effect; in many practical situations, inhomogeneous broadening will wash out the hyperfine structure of the spectrum. In such cases of inhomogeneously broadened EMR spectra, interaction can only be deconvoluted for practical analysis via simulations (Hyde & Froncisz, 1982). The ENDOR method, however, records a spectrum that represents the hyperfine structure alone, and the spectral lines thus observed are inherently selective and more narrow than the parent EMR spectrum. Other advantages of ENDOR that affect the analysis of complicated electron nuclear interactions, such as the factoring of multiple lines, are covered in other reviews (Kevan & Kispert, 1976; Schweiger, 1982).

Despite the technique's advantages, ENDOR can be problematic in the practical sense. The ENDOR signal amplitude depends on the spin–lattice relaxation rates that must be exceeded in order to observe an enhancement of the saturated EMR signal, and any physical phenomenon that accelerates relaxation rates is a bane to ENDOR spectroscopists. The so-called "ENDOR enhancement" (proportion of original EMR signal recovered by sweeping through the NMR transition) is strongly affected by such physical factors as the temperature and the viscosity (liquid) or lattice structure (solid) of the medium. Other commonly encountered rate-accelerating mechanisms that are inherent to the paramagnet itself might be dipole–dipole interactions of type electron–electron (exchange interactions; Anderson & Weiss, 1953) or nuclear–nuclear (cross-relaxation; Grant, 1964a–d; Standley & Vaughan, 1966; Verstelle, 1968).

4.2. ESEEM

Electron spin echo modulation spectroscopy (Norris *et al.*, 1980; Dikanov & Tsvetkov, 1992) is sometimes called FT-ENDOR because the echo modulation time series yields a frequency spectrum that corresponds to transitions among nuclear sublevel (Rowan *et al.*, 1965). The ESEEM technique is often said to be complementary to ENDOR (Tsvetkov & Dikanov, 1987) because ESEEM tends to yield well-resolved spectra in the low-frequency range (≤ 4 MHz) of the nuclear hyperfine spectrum, where cw-ENDOR is often problematic. The converse is likewise true: ESEEM tends to be problematic at recording hyperfine frequencies above 10 MHz.

ESEEM spectroscopy is based on the spin echo phenomenon first described by Hahn (1950). Two or more saturating microwave pulses are applied at or near the resonance condition with a sample and thereby shift the net magnetization of a sample by altering the relative populations of the spin ground and excited states. The two common pulse sequences are comprised of 2- and 3-pulses, which formally correspond to $\frac{1}{2}\pi - \tau - \pi$ and $\frac{1}{2}\pi - \tau - \frac{1}{2}\pi - T - \frac{1}{2}\pi$ excitation in the parlance of spin turning angles (*cf.* Freeman, 1997a;b), but pulses not exactly corresponding to the specified turning angles are acceptable (Hahn, 1950; Rowan *et al.*, 1965). The two pulse sequences differ with respect to the echo's response to spin–lattice relaxation and the manner of mixing two or more frequency components in the modulation pattern. In general, the three-pulse, or stimulated echo, technique is preferred because the echo amplitude persists for longer interpulse spacings and linear mixing terms in the spectrum are reduced in number, thus simplifying the interpretation of a multicomponent spectrum (Mims, 1968, 1972a).

The echo method is valuable because it provides a second means of determining spin relaxation rates that is an alternative to spin induction decay. The amplitude of the echo decreases monotonically as the interpulse spacing increases, and one can record an echo amplitude "envelope" as a discrete time series. Such a plot of the echo amplitude as a function of interpulse spacing can be fit to a decay functions (*cf.* Carr & Purcell, 1954) and be used to recover spin–lattice relaxation times, T_1, that might otherwise be measured by direct detection time domain EMR (see Kevan & Schwartz, 1979; Dalton, 1985). Figure 4a schematically illustrates the method for a simplified two-level system, drawing the analogy between fluorescence decay and spin echo decay.

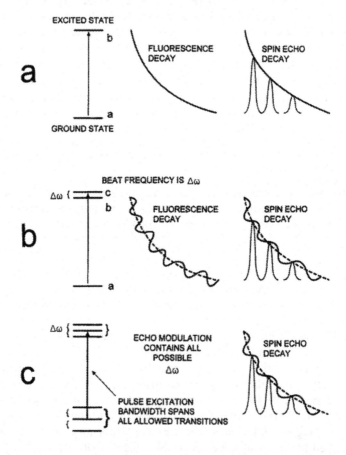

Figure 4. Excited state decay schemes from two- and three-level systems (a, and b, respectively), compared as measured using fluorescence spectroscopy and spin echo methods. When measuring decay from a single excited state to the ground state, a fluorescence experiment detects a single wavelength and decaying intensity; magnetic resonance, a spin echo of decreasing amplitude (**a**). Decay from two simultaneously driven excited states yields an oscillating intensity (fluorescence) or modulation of the spin echo amplitude – the so-called quantum beats phenomenon (**b**). In pulsed EMR the bandwidth of the high power pulse simultaneously drives all transitions of the six level diagram, the familiar beat pattern of weakly coupled ^{14}N (**c**).

If one modifies the state diagram by adding a third level, one could, in principle, measure two relaxation times (corresponding to the two radiation-induced transitions $a \rightarrow b$ and $a \rightarrow c$, Figure 4b) independently. One merely needs to ensure that the excitation source of one's spectrometer can independently select the two transitions, that is, the excitation source bandwidth is significantly less that the difference between the two transition energies. Decay from excited states b or c might be measured as a fluorescence lifetime or echo decay, depending on the nature of the system.

As the difference between the two transition energies becomes less than the bandwidth of the source, the experiment can no longer be used to distinguish processes $a \rightarrow b$ and $a \rightarrow c$, and the corresponding decays from the excited states begin to interfere classically, as would be described by light waves (*cf.* Michelson, 1903). This interference is manifest as a beat pattern that is superimposed onto the decay envelope, and the beat frequency corresponds (in the three-level model) to the energy difference between levels b and c, or, to put it more accurately, the difference between the transition energies of processes $a \rightarrow b$ and $a \rightarrow c$ (Alexandrov, 1964).

The model is examined using a simple EMR state diagram, $S=\frac{1}{2}$, $I=\frac{1}{2}$, in Figure 5, and the echo modulation process may be described as follows. The splitting between the states (m_S, m_I) $|+\frac{1}{2},-\frac{1}{2}\rangle$ and $|+\frac{1}{2},+\frac{1}{2}\rangle$ is determined by the

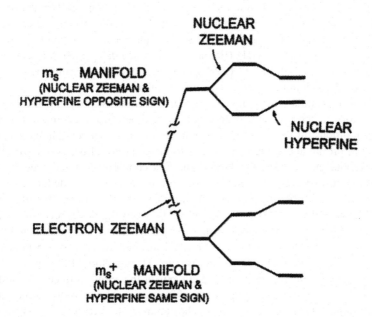

Figure 5. Four-level hyperfine diagram illustrating manifold with sign of nuclear hyperfine coupling opposite that of nuclear Zeeman term, making it possible to adjust the Zeeman field so that the spin manifold levels cross.

nuclear Zeeman and contact hyperfine terms. The nuclear Zeeman term, $g_n\beta_n H_0$, is determined by the applied field, and the splitting can therefore be experimentally controlled. If Fermi contact interaction a is positive, the spin quantum numbers render the signs of the nuclear Zeeman and contact terms opposite, and the levels corresponding to the states $|+\frac{1}{2}, -\frac{1}{2}\rangle$ and $|+\frac{1}{2}, +\frac{1}{2}\rangle$ can be made to cross by varying H_0. The states $|-\frac{1}{2}, -\frac{1}{2}\rangle$ and $|-\frac{1}{2}, +\frac{1}{2}\rangle$ cannot be made to cross because the spin quantum numbers are the same sign; these latter states therefore diverge in a linear fashion as the applied field increases, but one should recognize here that the field can be used to experimentally manipulate and "tune" the transition frequencies. This condition and process of forcing the near-degeneracy of the transition energies is called "exact cancellation" (Mims & Peisach, 1976; Singel, 1989).

EMR transitions $|-\frac{1}{2}, -\frac{1}{2}\rangle \rightarrow |+\frac{1}{2}, -\frac{1}{2}\rangle$ and $|-\frac{1}{2}, +\frac{1}{2}\rangle \rightarrow |+\frac{1}{2}, +\frac{1}{2}\rangle$ may be simultaneously driven by incident microwave field H_1 if the energy difference of these transitions is less than the spectrometer bandwidth, which is affected by the modulation or excitation method. In such a case quantum beats or modulation effects will be detected on the decay profile. For example, a 20-ns square pulse will have an excitation bandwidth of approximately 50 MHz, which is quite broad relative to the EMR linewidths of many organic radicals, and one might expect to detect the entire ENDOR spectrum of an organic radical in an echo modulation time series. The state diagram of the proton hyperfine (*i.e.,* ENDOR, Bender *et al.,* 1989) spectrum of ribonucleotide reductase is readily encompassed by the 65-MHz pulse bandwidth of a standard high-power pulsed EMR spectrometer, yet the ESEEM and cw-ENDOR spectra of ribonucleotide reductase do not in any way resemble one another at 9.5 GHz (Bender, unpublished data). Such marked disparities between cw-ENDOR and ESEEM spectra are generally the rule and one of the reasons that the techniques are called complementary.

The echo modulation frequencies are predicted classically, and the formulae are analogous to those found in descriptions of diode mixer circuits (*cf.* Carson, 1990). Conditions for the echo phenomenon are twofold, namely, inhomogeneity in the spectral profile and some nonlinearity to the response of the system (*cf.* Hahn, 1950; Chebotayev & Dubetsky, 1983). The first condition is readily met by most samples or spectrometers (Ramsey, 1950), and the latter may be deduced from the Bloch equations. Mims' descriptive model (1968, 1972a–c) of the modulation effect relies on a classical inductive coupling between dipoles (precessing gyros). The electron gyroscope precesses with an angular velocity is approximately 1000 times that of the nucleus. If the electron and nuclear gyroscopes are inductively coupled one can imagine that the otherwise circular trajectory of the electron's precession is made elliptical, and that the long axis of the ellipse (directed towards the nucleus) oscillates at the nuclear precession frequency. The echo that is recorded at 2τ (Hahn echo) or $2\tau + T$ (stimulated echo) amounts to a temporal "snapshot" of the primary magnetization vector, which is affected by both static field H_0 and the precessing magnetic moments of the local nuclei, and the time series that is generated by varying τ or T is therefore an interferogram of the nuclear precession frequencies (Hahn & Maxwell, 1951). In other words, the precessing electron, whose magnetization vector serves as an observable, acts as a linear mixer of these

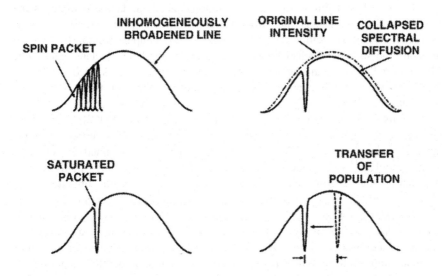

Figure 6. Spectral diffusion and its role in ENDOR and ESEEM. Top left: The inhomogene-ously broadened EMR line is represented by a superposition of (homogeneously broadened) spin packets that may be saturated so that a hole is burned in the line (Bottom Left). Spectral diffusion spreads the saturating power through portions of the spectrum and may collapse all or part of the EMR line (Top Right). In an ENDOR experiment, the population of a second spin packet is transferred to the portion of the spectrum where a hole has been burnt, and the whole is filled (recovery of saturated EMR signal). Cross-relaxation, via mutual dipole flips, fills the hole without the drive of the second rf frequency. This spontaneous hole-filling runs counter to the desired spin dynamics of a cw-ENDOR experiment, but provides the condi-tions to detect quantum beats in the decay via ESEEM.

nuclear frequencies by virtue of magnetic dipole coupling among the collective magnetic moments.

4.3. ENDOR vs. ESEEM: Experimental Technique

The need to drive spin transitions sufficiently fast to overcome the tendency of the relaxation rate to equilibrate spin state populations compromises the ENDOR technique. For example, the optimal rf drive conditions for observing protons of different type (*i.e.,* $\alpha-$, $\beta-$, $\gamma-$, *etc.*) by the ENDOR method vary (*cf.* Hyde *et al.,* 1968). Similarly, ENDOR transitions of nuclei whose gyromagnetic ratio is low tend to be inherently more difficult to observe (*e.g.,* ^{2}H, as opposed to ^{1}H). Such ENDOR experiments often require large H_2 fields, and the spectral lines tend to be relatively broad (Kevan & Kispert, 1976).

ENDOR is also very difficult to observe when cross-relaxation becomes a fac-tor and, in effect, "short-circuits" the relaxation routes, thus preventing observation of ENDOR enhancement. Often cited in the context of the complementary nature of ESEEM and ENDOR is the fact that among metal imidazole complexes the ENDOR technique detects the strongly coupled (metal-coordinated) ^{14}N only, whereas the ESEEM method detects only the weakly coupled (remote) ^{14}N. As it turns out, the six-level nuclear hyperfine diagram of the latter nucleus tends to be subject to a level-crossing condition at the DC magnetic fields used in X-band EMR experiments. This near level-crossing condition opens up the possibility of cross-relaxation (*e.g.,* via dipolar interaction, *cf.* Abragam & Proctor, 1958) and is therefore likely the mechanism responsible for the lack of an ENDOR spectrum for the weakly coupled imidazole nitrogen atom. This conclusion is supported by the fact that high-quality ENDOR and Triple spectra may be obtained for similarly weakly coupled ^{14}N in single-crystal samples of Cu(II)/Zn(II) *bis*-(diethyl-dithio-carbamate); the narrow lines of the single-crystal spectrum obviate cross-relaxation, and otherwise intractable spectra are observed (Böttcher *et al.*, 1984).

ESEEM, by contrast, is a form of coherence-transfer spectroscopy (Macomber, 1976; Hollas, 1982) that seems to perform optimally when the cross-relaxation phenomenon allows an admixture of states. The degree to which nuclear manifolds are mixed appears to be linked to the so-called "modulation depth," or peak–peak amplitude oscillations in the echo modulation time series relative to the average echo amplitude (again, the much-desired "exact cancellation" condition). The modulation depth, in turn, determines how sharp and well-defined spectral peaks are recovered in the Fourier transform, as is the case of any transformation of an oscillatory waveform into the frequency domain. Among the body of literature (Dikanov & Tsvetkov, 1992) for nuclei of similar gyromagnetic ratio and (pre-sumably) quadrupolar relaxation rates, there is a large disparity in ESEEM modula-tion depth, and therefore relaxation enhancement by quadrupolar effects (*cf.* Abragam, 1961) cannot be the sole factor responsible for deep electron spin echo modulation. The directly coordinated nitrogen will relax differently than the remote nitrogen in a metal imidazole because of the proximity of the metal, but if one is to invoke accelerated relaxation as a factor in modulation depth, then the coupling to the paramagnetic copper should further enhance the relaxation of the ligand nitro-gen and render the nucleus subject to detection by ESEEM. Similarly, the relaxa-tion rates of weakly coupled deuterons in, for example, a tyrosyl radical (Warnecke *et al.*, 1995) should not differ markedly from weakly coupled 2H$_2$O (Mims *et al.*, 1984), yet electron spin echoes are modulated much more deeply by the latter than the former. Although perhaps coincidental, it seems that the ESEEM technique is optimal when the energy levels of one spin manifold are sufficiently close to allow for cross-relaxation and admixture of these close-proximity states; under such con-ditions, the spectroscopic technique becomes analogous to coherence transfer and fluorescence enhancement methods in atomic and molecular spectroscopy, which also owe their narrow lineshape and intensity to a condition of level-crossing.

ESEEM spectra become, in general, inferior to cw-ENDOR spectra when the modulation depth becomes shallow and renders the FT-spectrum poorly resolved. This occurs as the otherwise admixed energy states disperse and "turn off" the cross-relaxation, and at this point conventional ENDOR becomes feasible. For ex-ample, excellent ^{2}H-ENDOR have been recorded for organic radicals, and the NQI tensors applied to structural analyses and the detection of phase transitions (Krzytek & Kwiram, 1991; Krzytek *et al.*, 1994, 1995). Another limitation of ESEEM relates to sampling theory and the manner in which the echo modulation is measured. The spin echo is typically a bell-shaped waveform of width approximately 15–20 ns, although the width varies as a function of T_2. The echo amplitude, which is plotted in order to generate the modulation time series, is recorded as an integrated area sampled with a gated amplifier whose sampling aperture approximates the echo width. Digital delays and time interval measurements can be performed on the subnanosecond time scale, but the inherent echo width and data acquisition method limit the timing resolution to 1 ns.

Besides the spin dynamics of the measurement itself, waveform sampling in the time domain for frequency analysis via transform methods are generally subjected to constraints such as the Nyquist theorem (1946), which stipulates that a waveform with highest frequency component f_m and duration T requires greater than $2Tf_m$ discrete points for accurate sampling and spectral representation via transform methods. In practice, this theorem stipulates that the waveform should be sampled in such a manner that there are more than two points per shortest period. A typical spectrometer may record an echo modulation time series as 1000 discrete points. Arbitrarily setting the number of sample points per shortest period as four, that is $1000 = 4Tf_m' > 2Tf_m$, the theoretical maximum frequency that may be resolved in the time series is f_m' (MHz) = 250/T (s). Hypothetically performing a representative ESEEM experiment by incrementing the interpulse spacing by 10 ns, a waveform of duration $T = 10\,\mu s$ will be generated with a Nyquist limit of $f_m' = 25$ MHz. The periods of, for example, 4- and 25-MHz modulation frequencies are 250 and 40 ns, respectively. The 15-ns gated amplifier aperture therefore represents 6 and 37% of the period in the respective cases, and the practical question of just how "discrete" are the sample points (exclusive of instrumental factors, waveform damping, and noise) becomes an issue. This limitation, combined with the dispersion of states excited by the finite frequency bandwidth of the excitation pulse, renders the ESEEM method specious for accurately recording modulation frequencies above 10 MHz; the situation improves if the echo (and the requisite amplifier aperture) is narrow. ENDOR and ESEEM, however, can be performed on the same spectrometer by using echo-detected swept frequency or field methods (Clark *et al.*, 1996), and a contiguous spectrum can be obtained by performing orchestrated experiments. Another option, however, is the coherent Raman beats method (Bowers & Mims 1959; Brewer & Hahn, 1973; Bowman, 1992), in which the modulation is detected as a "beat" against a continuous wave (or long pulse) signal. In this case the limiting factor in waveform recovery is the bandwidth of the receiver; waveform sampling oscilloscopes are, in effect, boxcar averagers and typically feature aperture times on the order of 100 ps.

5. ZEEMAN DEPENDENCE OF HYPERFINE SPECTRA

Thus far, we have established that the nuclear hyperfine parameters of the spin Hamiltonian are desirable for assessing the chemically interesting problem of structure–function correlation and reaction control. The advanced EMR methods known as ENDOR and ESEEM best recover this information from samples in which the chemical agent of interest is paramagnetic, and, in principle, there are methods that enable the spectroscopist to cope with the sometimes pathological behavior of spin systems, in other word, coax a spectrum out of a sample. In this section, however, we shall address the question of whether there is necessary and sufficient information in a single ENDOR or ESEEM spectrum and how to design an experimental approach that enables one to fully parameterize the spin Hamiltonian.

The spin Hamiltonian (eq. (2)) contains two first-order perturbation terms that are dependent upon the magnitude of applied field H_0, which is an experimentally controlled parameter. The electron magnetic resonance spectroscopy experiment can therefore be tailored in the sense that the hyperfine splitting between energy levels can be arbitrarily adjusted by varying the DC magnetic field as long as the fundamental electron resonance relation, $h\nu = g_e\beta_e H_0$, is maintained. This ability to control hyperfine splitting is important because one can thereby force the state diagram to assume configurations that greatly simplify the spectroscopy experiment and recovery of the desired molecular parameters.

Proton ENDOR transition frequencies, such as those that appear in Figure 3, are very nearly symmetrically dispersed about the proton Larmor frequency according to the empirical rule $\nu_\pm = \pm a$, where $\nu_\pm$ refers to the upper and lower frequency ENDOR transitions, and ν_n corresponds to the nuclear Larmor frequency at the experimental DC magnetic field (for ^1H, approximately 15 MHz at 0.35T). When $\nu_n > a$, the ENDOR transition frequencies assume positions in the spectrum according to the rule $\nu_\pm = \pm a\nu_n$. It follows that the peaks of an ENDOR experiment will shift as the experiment is repeated at different field/frequency combinations. For example, a proton ENDOR spectrum (isotropic a) will behave as illustrated in Figure 7. The reader should recognize that the plot depicted in Figure 7 represents the trajectory mapped out by ENDOR peak positions in discrete experiments; both spectrometer operating frequency and field are varied in order to maintain a constant g-value.

For experiments conducted at H_0 such that $\nu_n > a/2$, one observes that paired ENDOR transitions $\nu_\pm$ follow parallel trajectories that are separated by a. As one decreases H_0, ENDOR transition ν_- approaches zero as $\nu_n \sim a/2$, and then reverses direction as $\nu_n < a/2$. At the zero Zeeman field limit (*i.e.*, $\nu_n = 0$), both ν_- and ν_+ converge at value $a/2$. The minimum that is traced out by the trajectory of ν_- corresponds to an energy level crossing as states $|+\frac{1}{2}, +\frac{1}{2}\rangle$ and $|+\frac{1}{2}, -\frac{1}{2}\rangle$ become degenerate. This resultant degeneracy is due to the fact that the signs of the nuclear Zeeman and contact terms are opposite, and the magnitude of H_0 ensures that nuclear Zeeman and contact terms cancel. (If contact term a is negative, then states $|+\frac{1}{2}, \pm\frac{1}{2}\rangle$ cross.) As the nuclear Larmor energy further decreases beyond this criti-

cal point, the energy levels diverge and the peak position shifts to a higher frequency.

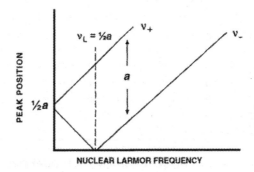

Figure 7. Zeeman dependence of ENDOR transitions in a S=1/2, I=1/2 spin system. As the
nuclear Zeeman energy is varied, the peak positions shift and there exists a critical point at
which the nuclear Zeeman energy equals half the nuclear hyperfine energy (dashed vertical
line). The critical point corresponds to the energy level (anti-)crossing situation.

For a simple model in which the spectroscopic energy terms are treated as scalar quantities and linewidths are represented by δ-functions, the trajectory of v_- is
V-shaped and the predicted line intensity will be uniform over the entire experimental frequency range. But for a real system in which there are finite linewidths,
such as that illustrated in Figure 9, the energy states will be more disperse and
permit overlap for some range $v_n - a$ greater than zero. This overlap of the dipolar
broadened states will permit cross-relaxation (*cf.* Geschwind, 1967), and the v_-
ENDOR line intensity will markedly decrease as the lines increasingly overlap.
The trajectory of the ENDOR peaks will likewise be modified and presumably
broaden at the so-called critical point because of the increasing probability of state
mixing as the adjoining states converge. In other words, the lower V-shaped trajectory will become more parabolic near the minimum as the overlapping disperse
states increase the bounds on nuclear Zeeman energies that permit level crossing. A
similar drop in signal intensity will likewise occur for the same reason at the zero-
field limit where all four states converge.

When $I \geq 1$ the addition of quadrupole coupling renders the hyperfine splitting
patterns asymmetric. The m_I levels are unequally spaced (without the quadrupole
coupling all levels would be equally spaced irrespective of I), and one obtains more
than two ENDOR transitions. For example, when $I=1$ there are two $\Delta m_I = 1$ transitions per m_S spin manifold, and their Zeeman dependence behavior is illustrated
in Figure 8. In contrast to the $I = \frac{1}{2}$ case, the trajectories are split by nuclear quadrupole coupling, and the level-crossing condition may occur between two m_I sublevels (*e.g.*, between +1 and 0) or among all three, depending on the spectral dis-

persion of those m_I energy states (*i.e.,* relative magnitude of dipolar interaction energy and the zero-field nuclear quadrupole interaction splittings).

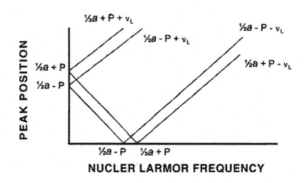

Figure 8. Level (anti-)crossing of nuclear sublevels S=1/2, I=1 by adjusting the nuclear Zeeman splitting. Only the m=1 transitions are shown in the diagram. As in Figure 7, the ENDOR transitions are mobile and subject to a critical point as the magnitude of the nuclear zeeman energy assumes a value that leads to the crossing condition in one electron spin manifold.

The (ENDOR) spectroscopic peak trajectories of Figures 7 and 8 illustrate how replicate experiments, that is, multi-frequency experiments conducted at the same *g*-value, result in identification of specific "critical points" characterized by spin-Hamiltonian parameters. These critical points correspond to level-crossing conditions that are imposed by changes in the nuclear Larmor component of the spin Hamiltonian as the DC magnetic field is varied. Figure 9 depicts the six-level state diagram ($S=\frac{1}{2}$, $I=1$) in which the individual levels are dispersed in energy because of dipolar effects and inhomogeneities (Geschwind, 1967).

5.1. Multi-Frequency Measurements and Energy Level-Crossing as an Interpretative Aid

Conventional EMR spectrometers operate at discrete frequencies within a given band. ^{1}H-ENDOR (and $S=\frac{1}{2}$, $I=1$ systems in general) are sufficiently simple that spectral interpretation can be routinely made at a single spectrometer operating frequency (*i.e.,* a single nuclear Larmor frequency on the plot in Figure 7) and, for instances in which spectra are highly convoluted, by using specialized procedures such as TRIPLE (Schweiger, 1982). But it is often necessary to interpret a multi-parameter ENDOR spectrum by numerical simulation of the spectrum using a spin-Hamiltonian model, and this becomes problematic because, in addition to the spin-Hamiltonian parameters for each nucleus, one also has to account for the relative orientations of the tensors. A multi-frequency approach to recording spectra is then called for in order to increase the number of spectroscopic observables (*i.e.*, peaks) in order to better refine the numerical simulations; the idea being that a correct assignment of spin-Hamiltonian parameters will enable one to simulate the spectra recorded at all spectrometer operating frequencies.

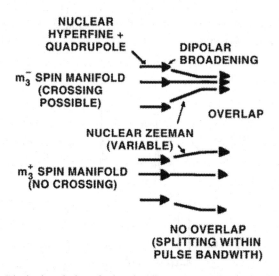

Figure 9. Dipolar broadening of otherwise discrete energy levels, leading to overlap and mixing in one spin manifold. This leads to cross-relaxation (ENDOR) and quantum beats (ESEEM).

What is typically termed multi-frequency EMR therefore entails the repetition of a spectroscopic experiment at a single frequency in the S-, C-, X-, K-, and Q-bands; the hyperfine parameters would then be culled from the data by independent simulations of the spectra (*cf.* Hyde & Froncisz, 1982; Hoffman *et al.*, 1993). In this experimental scenario, state degeneracy, whether as a crossing of states attributed to the same nucleus, as depicted in Figure 7, or between two or more nuclei, is something to be avoided because it becomes a factor in determining spin-relaxation rates and the radiofrequency power that becomes necessary in order to observe an ENDOR signal (Kevan & Kispert, 1976; Schweiger, 1982). In other words, for a cw-ENDOR experiment, the traditional multi-frequency approach to spectroscopic interpretation does not make use of the so-called critical points identified in Figures 7 and 8 because of the associated deleterious relaxation effects on the spectra.

Figure 10 illustrates a simulated set of three $S = \frac{1}{2}$, $I = 1$ hyperfine spectra performed in the single spectrum per band approach. The hyperfine parameters that were used to generate these data represent those of a single weakly coupled nitrogen, and even this simple example results in very complicated spectra whose peaks are difficult to assign. Furthermore, the only way to analyze this set of spectra and correctly determine the spin-Hamiltonian parameters would be to conduct numerous trial-and-error simulations in order to best fit the spectra using a set of spin-Hamiltonian terms and tensor orientations. This example graphically shows that, as simple as ENDOR can be for $I = \frac{1}{2}$ systems, the multi-frequency approach, as practiced in the manner described in the preceding paragraph, is problematic with $I \geq 1$ systems. It likewise follows that the quantitative analysis of hyperfine parameters by fitting a Hamiltonian model to EMR spectra recorded at discrete frequencies in separate bands leaves one open to error in selecting parameters for optimizing the simulation's fidelity to experiment.

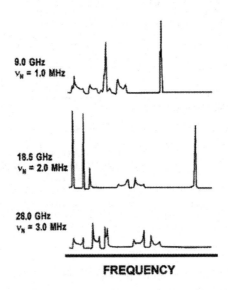

FREQUENCY

Figure 10. The conventional multi-frequency approach to ENDOR/ESEEM by recording spectra at discrete spectrometer operating frequencies in two or more microwave ovtaves. These data represent simulated ESEEM/ENDOR spectra of an S=1/2, I=1 system using the hyperrfine parameters e^2Qq=1.6 MHz, η=0.45, and A_{iso} = 4.0 MHz. Top and bottom spectra correspond to nuclear Zeeman energies above and below the ideal 'exact cancellation' condition (center spectrum). The simplified exact cancellation spectrum makes it easy to assign peaks to transitions (cf. Mims & Peisach, 1978), but peak mobility makes it difficult to assign numerical values to the hyperfine parameters based on a single spectrum.

Besides the ambiguity of fitting simulated to experimental spectra, the instrumental limits (most conventional microwave sources — klystrons, Gunn diodes, etc. — operate only a limited band width within the octave, that is, they are not tunable) of the "one frequency per octave" approach does not assure one of being able to obtain the requisite number of spectra to begin with. For example, it is highly problematic to record cw-ENDOR of weakly coupled ^{14}N at the X-band (because of cross-relaxation), and so the most common experimental configuration will not yield data of suitable quality so as to perform a meaningful Hammett-like analysis of the system. Instead, I propose that a more quantitatively reliable procedure may be formulated by using the Zeeman-dependent behavior of the hyperfine spectrum to provide a graphical approach to determining the spin Hamiltonian parameters (*cf.* Singel, 1989). The proposed multi-frequency approach will be used to record spectra in the region of so-called "exact cancellation" both to extract the spin-Hamiltonian parameters from the aforementioned critical points and to take advantage of line-narrowing phenomena that occur in ESEEM spectra near exact cancellation.

Figures 7 and 8 have been used to identify useful trends, namely, that level crossing in one m_S spin manifold causes the trajectory of at least one ENDOR transition frequency, as mapped by plotting peak position against the nuclear Zeeman energy, to have a minimum value (*i.e.,* a critical point) at which the hyperfine interaction energy may be read directly from the raw data. For the $S = \frac{1}{2}, I = \frac{1}{2}$ system this minimum corresponds to the contact interaction energy, but when $I \geq 1$ this minimum can be used to assign the zero-field NQI parameters. The improved multi-frequency experimental protocol therefore entails replicate measures of spectra at small increments of spectrometer operating frequency (500-MHz steps are ideal) and plotting peak position *vs.* the nuclear Larmor frequency. The use of small incremental steps allows one to, in effect, map the hyperfine spectrum energy levels and locate the level-crossing condition. This approach does not obviate the need for spectral simulations, but it does provide a higher level of confidence in the interpretation when one finally does try to correlate the experimental data with a spin Hamiltonian model and simulations. In short, the method obviates interpretive errors that might be associated with drastic lineshape changes that accompany large nuclear Zeeman energy steps (*i.e.,* Figure 10). As it turns out, the forced condition of level crossing also introduces nonlinear effects that yield strikingly narrow lines that enhance resolution of the quantitative measurement.

5.2. Level-Crossing Spectroscopy

The complementary nature of cw-ENDOR and ESEEM (or FT-ENDOR, to resurrect an old descriptive term; Rowan *et al.*, 1965) was described in §4 by referring to the differences between the role of cross-relaxation (spectral diffusion) in each of the two methods, and these cross-relaxation effects establish a connection between ESEEM and what is known as level-crossing spectroscopy. In general, level-crossing spectroscopy (Hollas, 1982) is identified with a situation in which the experimentalist controls the state diagram of the system under study, with the result that the spectrum is modified in some manner. This behavior was first associated with atomic fluorescence spectroscopy in which a marked enhancement of signal intensity and linewidth was obtained when the Zeeman component of the state energy was used to force crossing of the excited states (Franken, 1961; Cosgrove *et al.*, 1959). With the crossing of the excited states the transition energies from two or more ground states became equivalent, and in this respect the similarity between the experimental scenarios of atomic level-crossing fluorescence and ESEEM spectroscopies becomes apparent.

With a nominal pulse width of 20 ns, the bandwidth of a conventional high-power pulsed EMR spectrometer is approximately 50 MHz, which would readily encompass the dispersion of allowed $\Delta m_S = 1$ transitions in Figure 2. At the same time, the Zeeman field has been chosen so as to force the merger of the m_I states of one electron spin manifold, and therefore the microwave pulse is driving multiple ground states into a common admixed excited state in precise analogy to the atomic fluorescence case. ESEEM spectroscopy is therefore a form of "level-crossing" spectroscopy applied to nuclear hyperfine spectra and should be generally applicable to those systems in which the nuclear sublevels of one m_S spin manifold "cross" and ground states simultaneously excited by an intense pulse of adequate bandwidth. For ENDOR spectroscopy, however, level-crossing is a catastrophic condition because of the associated relaxation effects that destroy signal amplitude.

5.2.1. Level (Anti-) Crossing and ENDOR

The first ramification of excited-state degeneracy is manifest in cw-ENDOR as what has to be the most frustrating aspect of the technique, namely, the balancing act between radiation-induced transitions and spin-relaxation rates that is necessary to observe an ENDOR signal. The cw-ENDOR signal intensity, which is often described as an "enhancement factor," depends on one's ability to drive the nuclear sublevel transitions faster than the electron spin–lattice relaxation rate. Transition probabilities and rates of population dynamics differ when the excited state of a spectroscopic transition becomes degenerate. For example, a very simple model system may entail a ground state a that is driven into a pair of degenerate levels b and c. Such a model describes the state–state transitions derived from resonance fluorescence experiments involving crossed atomic excited states. In such systems the observed enhancement in fluorescence yield (related to rate of energy disposal by radiative mechanism) is explained by the necessity of mixing states. The rate expression is therefore altered from a sum of individual transition rates. For example,

$$T_{\text{DISCRETE}}^{-1} \approx \left| f_{ab} g_{ba} \right|^2 + \left| f_{ac} g_{ca} \right|^2$$

is rewritten in a form that reflects the state admixture:

$$T_{\text{CROSSED}}^{-1} \approx \left| f_{ab} g_{bc} + f_{ac} g_{ca} \right|^2$$

Terms f_{ij} and g_{ji} represent expectation values (*e.g.,* oscillator strengths) of the excitation and relaxation process, respectively. It follows that, in general, the transition rate under the condition of crossing levels is greater than the discrete, which follows from Cauchy's theorem.

Level crossing and its effect upon spectroscopic transition rates in ENDOR are mechanistically understood in the context of linewidth and spin relaxation (Poole

& Farach, 1987; Caspers, 1964). Inhomogeneous broadening of a spectral line means that for a given transition $a \rightarrow b$, there is an associated transition energy $h\nu_{ab}$. But, because of inhomogeneities in the sample inhomogeneities in the sample field, there are populations of molecules whose $a \rightarrow b$ transition is driven above and below the "exact" resonance condition $h\nu_{ab}$. The individual sets of localized molecules themselves constitute a set of (Gaussian or Lorentzian) homogeneous lines that overlap to some extent (Grant, 1964a–d; Geschwind, 1967). This overlap determines the extent of "crosstalk" between the individual homogeneous line packets, and energy is disposed (relaxation) via spectral diffusion through the inhomogeneously broadened line via dipolar interactions. Linewidth broadening disperses the discrete energy levels of Figure 2, and there is a greater chance of overlap, mixing, and enhancement of cross-relaxation rates as the nuclear Zeeman energy brings the otherwise discrete levels into close proximity (Figure 9). In the circuit analogue diagrams that are used to describe the ENDOR phenomenon, this spectral dispersion has the effect of introducing additional low-resistance routes to the network, and one has to drive the "source" (the allowed EMR transition) much harder. Representative examples of level crossing and the effect on ENDOR line intensities in F-centers (Doyle & Dutton, 1969) may be compared with a study of cross-relaxation in ruby (Squire, 1965).

5.2.2. Level (Anti-) Crossing and ESEEM

In the preceding section, level crossing among magnetic resonance hyperfine states was described as a mechanism that affects the transition rates in a conventional EMR experiment, namely, swept frequency generation of the spectrum and low powers. An ESEEM or true FT-EMR experiment differs from the conventional technique in the sense that the powers that are applied to the sample are very intense (typically 20–25 W) and the time interval over which the resonant radiation is applied is very fast on the time scale of the spin relaxation rate. The microwave pulse widths, which are used in many experiments, also satisfy condition $t_p \leq 1/\Delta\omega$, where $\Delta\omega$ is the frequency dispersion of the microwave source, and one therefore observes nonlinear behavior among the spectroscopic states in the system (Klauder & Sudarsham, 1968); this nonlinearity is one of the necessary conditions for observing an echo (Hahn, 1950; Chebotayev & Dubetsky, 1983).

In the introductory paragraphs of this section it was stated that the pulse excitation bandwidth spanned the ground state energy dispersion of all the allowed transitions in a typical $S = \frac{1}{2}$, $I = 1$ hyperfine spectrum and that the excited states could be put near a crossing condition by manipulating the Zeeman field. For example, the ^{14}N nuclear Zeeman energy at $0.35T$ is approximately equal to the weak hyperfine interaction energy of remote nitrogen atoms in metal complexes, such as the non-coordinated imidazole ring nitrogen or the nitrogen atoms of diethyl-dithiocarbamate, so that the three m_I states (here regarding an $I = 1$ hyperfine interaction) become nearly degenerate. In other words, the energy levels can be induced to cross, and the corresponding states mix in one m_S spin manifold. The term "exact cancellation" was originally coined and meant to describe the phenomenon as a condition in which the nuclear Zeeman and contact terms were of approximately equal and opposite sign, leaving only the nuclear quadrupole term of the perturbational spin Hamiltonian, which led to the analogy between the ESEEM FT-spectrum and ZF-NQR spectroscopy (Mims & Peisach, 1978). The ESEEM FT-spectra of metal–imidazole complexes (most successfully Cu(II) and low spin Fe(III), which are ground-state $S = \frac{1}{2}$ ions) resemble the nitrogen ZF-NQR spectra of pure imidazole (>N–H fragment) with two intense features at low frequency (*ca.* 0.7 and 1.5 MHz) that correspond to ZF-NQR and Townes-Dailey (1955) frequencies $\nu_\pm = \frac{3}{4}e^2 Qq_{zz}[1 \pm \frac{1}{3}\eta]$ and $\nu_0 = \frac{1}{2}e^2 Qq_{zz}$. More detailed theoretical descriptions of the exact cancellation effect and the transition frequencies of the $S = \frac{1}{2}$, $I = 1$ system (Shubin & Dikanov, 1983; Iwasaki *et al.*, 1986) concur with the Mims (1972b,c) density matrix formalism and the general theory of quantum beats (Alexandrov, 1964).

Among ESEEM spectra of weakly coupled ^{14}N nuclei, the exact cancellation spectrum consists of three intense and narrow lines in the frequency range 0.5–2 MHz (Figure 11, top). For the case where $\eta \approx 1$, which seems to correspond to most metal–imidazole complexes in frozen polar solvents, the best resolved peak is located at ~1.5 MHz and corresponds to the $\Delta m_I = 2$ transition ν_+, which in the zero-field Townes-Dailey (1955) formalism, and corresponds to $\frac{3}{4}e^2 Qq_{zz}[1+\eta]$. The other two lines corresponding to ν_- and ν_0 partially overlap in this case of large η, but the striking aspect of these peaks that arises from the (anti-)crossing energy levels is their narrow width (particularly evident at 1.5 MHz), which is on the order of 10 kHz and approaches the natural linewidth that might be recorded from a single-crystal ENDOR experiment. As one would expect, the intensity and characteristic width of a given peak in an ESEEM spectrum will reflect the depth of the modulation in the time-series data. This modulation depth seems to be associated with systems in which the Zeeman energy brings the excited state levels into close proximity rather than the quadrupolar relaxation rate of the nucleus, which is consistent with the level (anti-)crossing spectroscopy analogy and mechanisms of coherence transfer spectroscopic techniques.

5.2.3. Level (Anti-) Crossing and the Spin Hamiltonian

Quantum mechanics abhors an intersection of two (or more) energy levels, which are repulsed in a manner that may be described by the perturbation theory as is commonly found in textbooks on the quantum mechanics (Kauzmann, 1957; Greiner, 1993). For example, the two-level perturbation theory leads to a power series expansion of terms in H_{ii} and H_{ij} and may be modified for n-levels (*cf.* Corson, 1951):

$$W_1 = \frac{H_{11} + H_{12}H_{21}}{H_{11} - H_{22}}$$

and

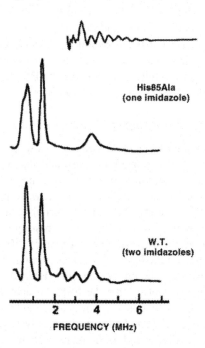

Figure 11. ESEEM spectra of the oxidized form of the Type I copper protein rusticyanin, which has the classic His₂CysMet Cu(II) binding motif. These are exact cancellation-like spectra that are obtained at spectrometer operating frequencies from 7.0 to 13 GHz. The peaks are mobile, as predicted by the nuclear Zeeman dependence. The upper spectrum is derived from an engineered form of the protein in which one of the two ligand histidines is removed; the spectrum is characteristic of a single ^{14}N hyperfine spectrum. The lower spectrum corresponds to the modulation spectrum of two nearly equivalent ^{14}N interactions and features the sum/harmonic lines in the 2-3 MHz region.

$$W_2 = \mathbf{H}_{22} + \frac{\mathbf{H}_{12}\mathbf{H}_{21}}{\mathbf{H}_{11} - \mathbf{H}_{22}} + \cdots$$

where terms $\mathbf{H}_{ii}$ correspond to the spin Hamiltonian levels (eq. (1)). The cross terms arise from the mixing of the near-degenerate states. It is evident that the power-series expansion that appears in this simplified expression is only an approximation to the solution of the secular determinant because of the singularity that arises as the levels "cross," but alternatives to the power series expansion are solved in the general case and applicable to the crossing problem (Slater, 1968; Schweiger et al., 1979). These equations are intended only to illustrate that the conventional spin-Hamiltonian expansion of terms that are used to interpret

ESEEM spectra at "exact cancellation" are close approximations to the spectra because the terms H_{ii} are retained. It follows from this simple illustration that, although one must force a near-crossing condition in order to optimize the modulation effect, this near degeneracy does not preclude the interpretation of the "exact cancellation" spectra by using the conventional spin-Hamiltonian terms. In other words, cw- and FT-ENDOR are indeed analogous; the ENDOR crossing problem has been examined in depth by Schweiger *et al.* (1979).

5.2.4. Ramifications of Level Crossing in ESEEM

Assuming that the analogy to level-crossing spectroscopy holds, ESEEM can be described as a situation in which an admixed excited state $\psi = |m_S^-, +1\rangle + |m_S^-, 0\rangle + |m_S^-, -1\rangle$ is being pumped simultaneously from states $|m_S^+, +1\rangle, |m_S^+, 0\rangle$, and $|m_S^+, -1\rangle$ because the excitation bandwidth of the microwave pulse, which goes as $\Delta f = t_w^{-1}$, is greater than the separation between the ground-state energies. The hyperfine transition energies that arise from this (anti-)crossing condition are derived from the Zener and similar theories of state degeneracy lifting, and the narrow linewidths may be explained on the basis of this state mixing, which is the scenario of the enhancement observed in the optical systems (Breit, 1938; Geneaux *et al.*, 1969). In this regard, it is noteworthy that the ESEEM technique (almost universally performed at the X-band) often yields poor quality modulation time series and FT-spectra when the nucleus being probed is not a weakly coupled ^{14}N or ^{2}H. The quadrupole relaxation rate is not related to the relative contribution of the NQI energy to the total hyperfine interaction, and besides, the double quantum transition at 4 MHz is always less intense than the quadrupolar transitions from the opposite manifold despite the fact that the quadrupole relaxation components should be the same in both m_S manifolds. It therefore does not follow that the quadrupole relaxation of weakly coupled ^{14}N alone governs the echo modulation depth, and that instead, it is the cross-relaxation effects of the near crossing levels that determine the modulation depth and the quality of the FT-spectrum.

Analogies may be drawn between the ESEEM experiment and the Hanle effect, and therefore theories of interference fluorescence, quantum scattering, and the transition frequencies observed in these quantum beats experiments (*cf.* Hollas, 1982) are related and directly applicable to ESEEM. An interesting variation in the echo modulation phenomenon that demonstrates the relationship to quantum beats was demonstrated in NQR experiments (Gerishkin & Shishkin, 1973). Rather than simultaneously drive several states into a common excited state, as generally done in ESEEM, these NQR experiments entailed multiple transitions $1/2 \rightarrow 3/2$ and $3/2 \rightarrow 5/2$ that were simultaneously driven by two frequencies in the pulse sequence. In the zero-field condition, systems whose asymmetry parameter was zero did not exhibit beats, in contrast to systems in which $\eta > 0$; beats in the echo could be induced when $\eta = 0$ by imposing a Zeeman field, and this was interpreted as being caused by forced mixing of the magnetic states.

If ESEEM is indeed a variation of level (anti-)crossing spectroscopy, then is should be possible to perform ZF-NQR studies of any species in the region of the paramagnet provided that two conditions are met. The first of these conditions is that the energy levels (or states) be manipulated via the applied Zeeman field so that near degeneracy of hyperfine states occurs. This condition will depend on the form of the spin Hamiltonian (*i.e.*, whether terms with signs opposing that of the nuclear Zeeman term exist) and the magnitude of the dipolar broadening of the levels (dispersion of the states). In short, the states have to be able to be brought close enough for cross-relaxation to occur. The second condition requires that the separation of the ground state energy levels be less than the bandwidth of the microwave pulse that is driving the spin system. The power spectrum bandwidth, corresponding to $\Delta f \approx 50$ MHz for a 20-ns pulse, is a fairly large excitation band and should suffice for many types of electron–nuclear coupling ranges. The first condition is therefore likely to be the most important prerequisite for deep modulation and detection of effective ZF-NQI transitions, and it is noteworthy that recent high-frequency ESEEM have recorded electron echo modulation presumably by the strongly coupled ^{14}N nuclei of imidazole that are directly coordinated to copper, which, at the high fields used, approximately conform to condition 1 (Coremans *et al.*, 1997).

Finally, the level-crossing mechanism might be advantageously used to probe nuclei for which one or both of the aforementioned conditions may not be met. For example, anomalous spin relaxation was measured in nuclear systems at a prescribed field by Gutowsky and Woessner (1958). The anomalous relaxation occurs because two spin systems, ^{1}H and ^{35}Cl, and are brought into resonance by adjusting the Zeeman energies so that levels of dissimilar nuclei coincide (Figure 12). When the spin states are in resonance a flip-flop (cross-relaxation) mechanism of induced transitions via dipolar interactions may occur, and hence the enhanced relaxation rate.

This property might be used to advantage in an ESEEM experiment. The lower panel of Figure 12 depicts a hypothetical hyperfine pattern of the Gutowsky system. An arbitrary electron spin state, m_S, is first split by the zero-field term of the ^{35}Cl and "adjusted" to the resonance value at H_c. One then adds the contribution of the proton hyperfine splitting, and it is apparent that the resultant-line state description results in some closely packed states, two of which do cross at H_c. This type of behavior is the underlying mechanism of such techniques as coherence-transfer ENDOR (Höfer *et al.*, 1986; Mehring *et al.*, 1986), but in the hypothetical experiment described in Figure 12 ESEEM should yield exact cancellation like-modulation.

6. A GRAPHICAL APPROACH TO HYPERFINE SPECTRA ANALYSIS

The extent to which level crossing can be imposed upon a given system will depend upon the zero-field splitting of quadrupolar nuclei. For example, if the ZF-NQI splitting is larger than the dipolar broadening, it may not be possible to force overlap of the discrete energy levels. Nitrogen hyperfine studies are important

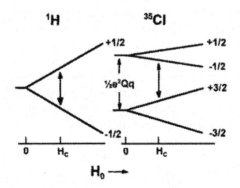

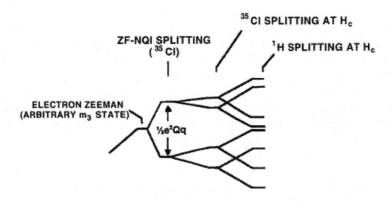

Figure 12. Cross-relaxation between different types of nuclei (^{1}H and ^{35}Cl) in NMR as the nuclear Zeeman levels cross (adapted with permission from Gutkowsky & Woesner, 1958). It is suggested in this chapter that the level-crossing phenomenon underlies the echo modulation effect, and the mechanism that yields cross-relaxation in the NMR experiment may be a valid means to detect modulation of nuclei whose own m_I levels cannot be brought sufficiently close to crossing. For example, ^{35}Cl are made to cross with ^{1}H levels by adjusting the Zeeman field strength (bottom state diagram).

because the nucleus is ubiquitous in both synthetic and natural metal complexes that are of chemical interest. It is fortuitous that the nuclear quadrupole coupling of many nitrogen compounds is less than 6 MHz (Lucken, 1969b), and therefore a level-crossing condition may be imposed even though the electron–nuclear dipolar interaction energy is weak. Boron, tin, and some oxygen nuclear quadrupole interaction parameters are also sufficiently small that the electron–nuclear dipole interaction energy may likewise lead to state admixture (*cf.* Lucken, 1969b; Schempp & Bray, 1970). Other nuclei whose nuclear quadrupole parameters are large, such as the halogens, sulfur, and most transition metals, may be induced into crossing conditions with other nuclei in order to create the desired effect. Although generally applicable, the principles of graphically analyzing hyperfine spectra will be illustrated for the $S=\frac{1}{2}$, $I=1$ spin system with ^{14}N parameters, for which ESEEM excels.

6.1. Peak Trajectories and Their Interpretation

The simulated 14N ESEEM spectra depicted in Figure 10 include all of the transitions of an $S=\frac{1}{2}$, $I=1$ spin system. These transitions are identified in Figure 13, which is a simulation of the spectrum at exact cancellation (*i.e.*, 18.5 GHz) using the same hyperfine parameters as was used to generate the spectra in Figure 10. But in Figure 13 a decay function, $\exp(t/T_2)^{\frac{1}{2}}$, where $T_2 = 2000$ ns (*cf.* Carr & Purcell, 1954; Klauder & Anderson, 1962), has been applied to the modulation time series prior to transformation, which results in a more realistic depiction of an experimental FT-spectrum and loss of resolution as the $\Delta m_I = 1$ transitions from m_S^+, in particular, are washed out. Two data sets are illustrated in the figure: one depicts the hyperfine spectrum of a single ^{14}N nucleus coupled to an electron spin, and the second depicts the interaction between two identical ^{14}N nuclei and the resultant combination lines (sums and harmonics).

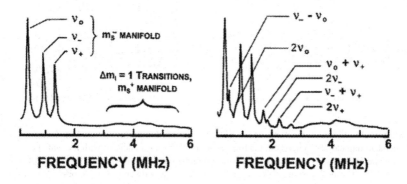

Figure 13. Simulated ESEEM spectra for the S=1/2, I=1 spin system depicted in the center trace of Figure 10. (left: one coupled nucleus; right: two coupled nuclei). A decay function has been applied to the modulation time series, and the FT-ENDOR spectra shown therefore exhibit a realistic lineshape. Peaks are labelled according to the scheme used in Figure 2.

The peak positions of the spectra illustrated in Figure 13 vary with the Zeeman field, and numerical assignments to spin Hamiltonian parameters based on inspection and a Townes-Dailey ZF-NQI interpretation will yield incorrect values of e^2Qq_{zz} and η. The exact cancellation-like spectral profile is retained over a wide span of microwave frequencies and one does not *a priori* know from a single ESEEM spectrum whether one is near true exact cancellation. In other words, one cannot simply tune up the spectrometer, locate a frequency/field combination that yields an exact cancellation-like spectrum, and interpret that spectrum using a Townes-Dailey or even the spin Hamiltonian model of eq. (1). And, as demonstrated in Figure 10, replicate experiments at separate microwave octaves still leaves one with a complicated analysis problem. One therefore needs a "continuously" tunable spectrometer to map the Zeeman dependence of the various peaks in the hyperfine spectrum.

The peaks of an ESEEM spectrum are Zeeman field dependent and may be used to develop graphical procedures for determining e_2Qq_{zz} and η (Flanagan & Singel, 1987; Flanagan et al., 1988; Singel, 1989). If one conducts a series of ESEEM experiments taken at small intervals in spectrometer operating frequency (e.g., 500 MHz), the changes in nuclear Zeeman energy alter the state diagram in commensurately small steps, and the peaks of the spectrum can be mapped so that they form a graphical "trajectory." Subject to the conditions described in the preceding section and this section's introductory paragraph, the energy levels of a hyperfine state diagram can be induced to cross, which is reflected in the trajectory of the peaks and used as an interpretative tool (cf. Figures 7 and 8).

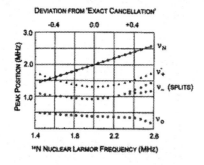
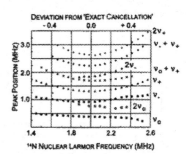

Figure 14. The peak positions of the ESEEM spectra of Figure 13 plotted as Zeeman dependent trajectories. When a≥P, the trajectories of the peaks associated with the (anti-) crossing energy levels are parabolas from whose minima one can directly read the ZF-NQR parameters. The peaks from non-crossing levels trace linear diverging trajectories. Note that the sum/harmonic (i.e. combination) lines mirror the fundamentals and can be distinguished from lines of other origin based on their behavior.

The peak positions of the simulated ^{14}N ESEEM spectra illustrated in Figure 13 are plotted as a function of the ^{14}N nuclear Zeeman energy in Figure 14, and these represent data recorded as discrete experiments in a manner analogous to that depicted in Figure 8. The peak trajectories are approximately parabolic when the associated transition corresponds to energy levels that "cross," and linear when the transitions are derived from levels that cannot cross. The minimum value of either type of plot yields the effective zero-field Hamiltonian parameters from which the ZF-NQI parameters can be culled. For a six-level state diagram such as the one depicted in Figure 2, the peak trajectories associated with the (crossing) m_S^- spin manifold are parabolic, whereas the m_S^+ peaks shift in a linear monotonically increasing fashion. The shape of the parabola's well would presumably reflect the dipolar broadening of the states; a broader dispersion of states increases the Zeeman field range that permits state admixture (Figure 9).

The trajectories of the peaks associated with the (non-crossing) m_S^+ spin manifold are linear monotonic increasing with the Zeeman field because all the terms of the spin Hamiltonian effectively add, and there is no nuclear Zeeman energy for which the three nuclear sublevels are close enough so that the states admix. It follows, therefore, that the trajectory of the peaks in general will reflect the zero-field splitting of the associated m_S electron spin manifold. For example, if $\mathbf{P}$ is significantly greater than a, then the zero-field splitting may be sufficiently large that no nuclear Zeeman energy will force level (anti-) crossing. The graphical method of analysis is therefore expounded as case studies that correspond to 1) levels that can be induced to cross, and 2) levels that do not cross.

The state diagram of the hyperfine spectroscopy experiment is controlled only to the extent that the nuclear Zeeman energy may used to null out other contributions to the hyperfine splitting. The nuclear quadrupole interaction is obtained as effective zero-field splitting values, and therefore the prospect of using the Zeeman field as a means to induce level (anti-)crossing will depend on the relative magnitude of $\mathbf{P}$ and a. If $a > \mathbf{P}$, then it is possible to generate a parabolic trajectory, but the parabola well flattens as $\mathbf{P}$ increases and ultimately assumes a linear trajectory as $\mathbf{P} > a$. Figure 15 illustrates the lineshape variations that arise as the relative magnitudes of spin-Hamiltonian parameters a and $\mathbf{P}$ are juxtaposed. In one plot $\mathbf{P}$ is varied for a fixed value of a, whereas the second plot illustrates the effect of varying a for a fixed value of $\mathbf{P}$. One can identify in Figure 15 each of the peaks that were assigned in Figure 13, and the Zeeman-dependent trajectories of the peaks are parabolic (Figure 14) so long as $a > \mathbf{P}$. Figure 16 illustrates the trajectory of the peaks when $\mathbf{P} > a$: because no value of $\nu_{^{14}N}$ can force crossing, the trajectories resemble a half-parabola in the sense that the plot assumes a minimum value below some critical value of nuclear Zeeman energy. It is evident from comparing the trajectory plots depicted in Figures 14 and 16 that combination lines (of a nucleus characterized by $a > \mathbf{P}$, e.g., imino nitrogen) may be distinguished from low-intensity peaks arising from a different class of the same nucleus ($\mathbf{P} > a$, e.g., amino fragment) simply on the basis of the Zeeman dependence.

6.2. Suppression Effects and Deconvolution of Lines

The graphical method of analysis works well only so long as the peaks are well resolved. Such an idealized situation might not be the case, however, and a technique for selective excitation is necessary for deconvoluting overlapping lines that may or may not have disparate origins. The density matrix model of the spin-Hamiltonian basis of echo modulation, however, features certain relationships between the timing parameters of the pulse sequence and the modulation frequencies that would enable one to effectively suppress peaks in the FT-spectrum (Mims, 1972a–c). For example, matching tau in the stimulated echo sequence to a frequency component of the spectrum will render that frequency transparent, and it will be lost from the modulation and corresponding FT-spectrum. This suppression effect may be used as a diagnostic tool in three-pulse ESEEM studies (Mercks & deBeer, 1979; Bender et al., 1997; Bender & Peisach, 1998) and is the basis of such modern pulse sequences as HYSCORE (Höfer, 1994).

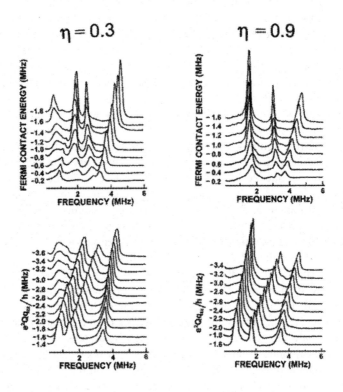

Figure 15. Variation of spectral lineshape as the character of the hyperfine spectrum changes from one in which a≥P to one in which P>a. Two extreme values of the asymmetry parameter η are assumed in each of these case studies.

A simple demonstration of tau-suppression is illustrated in Figure 17, which again uses the idealized model of a weak ^{14}N hyperfine spectrum. The top spectrum corresponds to data recorded with typical stimulated echo pulse parameters, specifically, tau equal to 150 ns, the starting value of the interpulse spacing between pulses 2 and 3 set to 70 ns, and the latter incremented by 10 ns. The complete spectrum is observed because tau does not correspond to any harmonic of the modulation frequencies extant. But, if one records the modulation series with tau set to 126 ns, which corresponds to the period of the $\Delta m_I = 2$ line at 7.94 MHz, there is observed a suppression of the (fundamental) peaks at 0.9, 1.4, and 8 MHz. Similarly, experiments with $\tau = 239$ ns and $\tau = 758$ ns collapse or enhance the fundamental lines. Note also that in each case the combination lines of the suppressed lines also collapse, and this can be used to deconvolute spectral peaks. For example, if there were a proton or second ^{14}N line in the region littered with combination lines, the combination lines could be suppressed in the manner demonstrated and the former identified and analyzed by the same graphical procedures as outlined above.

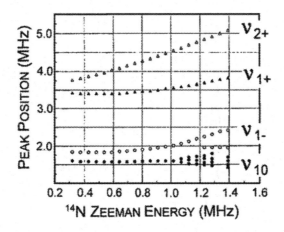

Figure 16. Zeeman field dependent trajectories of peak positions in ESEEM spectra when P>a. In this case the effective zero field splitting of the mI levels prevents crossing, and the peaks follow trajectories that resemble flattened half-parabolas. The behavior of this class of nuclei is easily distinguished from the a≥P class (and their combination lines).

6.3. Application of the Principles to a Strong Hyperfine Coupling

The graphical analysis of ENDOR or ESEEM spectral lines makes it possible to obtain ZF-NQI parameters for any quadrupolar nucleus provided that a series of spectra can be obtained as the magnetic field is incremented so that the levels of one spin manifold can be made to cross (*i.e.*, exact cancellation). This condition is easily imposed on weakly coupled ^{14}N nuclei (*i.e.*, the distal nitrogen of the imidazole/histidine ligands to copper), but our aim is to apply the technique to all hyperfine spectra, strong and weakly coupled nuclei, in order to make avail of all the probes potentially at our disposal in a given metal binding site. And this means that EMR spectrometers must operate at frequencies considerably higher than X-band (8–12 GHz).

Coremans *et al.* (1997) obtained ESEEM spectra for the HisN$_\delta$ of histidine that is coordinated to the copper ion in azurin. The profile of these spectra resemble those of the weakly coupled HisN$_\varepsilon$ (*i.e.,* non-coordinated) nitrogens, implying that, as expected, the nuclear Zeeman term of the spin Hamiltonian is sufficiently large at the W-band to render the energy levels close to the condition of exact cancellation. In this case $a \gg$ **P**, and it is desirable to know whether an incremental tuning of the spectrometer operating field and a graphical analysis of the resultant spectra would as easily recover the ZF-NQI parameters as has been demonstrated with weakly coupled nitrogen.

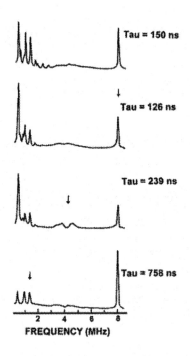

Figure 17. Tau-suppression effects in ESEEM spectra and their potential use in identifying and correlating peaks. In a semi-classical model, the procedure entails the locking of the temporal aspects of the pulse sequence, that is, the preparation-evolution-detection regions(see Ponti & Schweiger, 1994) to the precession frequency of a specific ENDOR resonance. One thereby renders the ENDOR transition transparent to the echo modulation interferogram.

The hyperfine splitting parameters of the two azurin HisN$_\delta$ were determined via multi-frequency cw-EMR studies by Antholine *et al.* (1993), yielding the following principal values for the hyperfine tensor (in MHz):

xx	yy	zz
27	25.5	27
21	18	17.5

For the pair, and the respective isotropic (contact) interactions of 26.5 and 18.8 MHz. The exact cancellation condition for the two nitrogen nuclei would therefore occur at $v = \frac{1}{2}a$, or 13 and 9.4 MHz, respectively, and the field required to promote the nuclear Zeeman energy to 13 MHz is $4.25T$, which is a 122-GHz experiment at $g = 2.05$. The corresponding field and spectrometer operating frequency that would cross levels of the more weakly coupled nitrogen is $3.07T$ (88 GHz). The predicted splitting of the $\Delta m_I = 2$ transition of the m_S^+ spin manifold at $4.2T$ is approximately 52 MHz, and the three ground states are therefore within the excitation bandwidth of a 15-ns microwave pulse, which is routinely achieved.

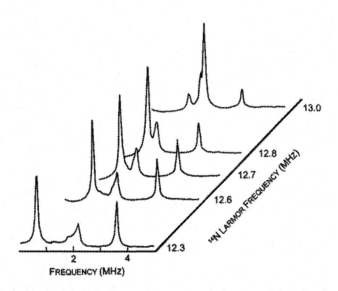

Figure 18. The simulated multifrequency ESEEM experiment conducted at W-band and using an S=1/2, I=1 spin system that models the strongly coupled hyperfine parameters attributed to the directly coordinated imidazole nitrogen of azurin, as determined by cw-EMR (Antholine et al., 1993). Note that in this model exact cancellation yields ZF-NQI like spectra whose peaks shift in a manner that is analogous to the weakly couples ^{14}N case.

The simulated spectra in Figure 18 represent a series of replicate ESEEM experiments performed on a system containing a single Cu–^{14}N interaction with a Fermi contact interaction energy of 26 MHz and quadrupole coupling ($e^2 Q q_{zz}$) of 3.5 MHz. It is immediately evident from these simulations that the behavior observed at the X-band is reproduced at the W-band provided that the spin-Hamiltonian model that is used to describe experiments at the X-band is not affected in the extremely high magnetic fields of the W-band (*i.e.,* no introduction of nonlinear effects). The W-band ESEEM spectra of the Type I copper center of azurin feature low-frequency ZF-NQI-like patterns (Coremans *et al.*, 1997), and so a hypothetical incremental study of the level-crossing experiment in the manner depicted in Figure 18 appears to be feasible for the graphical analysis of NQI parameters. The instrumental requirement, however, is that one have an incrementally tunable microwave source at the W-band so that the analysis can be performed reliably.

7. APPLICATION OF THE ANALYTICAL METHOD

The stated aim of this review is to demonstrate that classical analyses of physical organic chemistry are feasible with respect to complex systems such as supported metal catalysts through the application of advanced EMR spectroscopic techniques and determining the relevant spin Hamiltonian parameters via the Zeeman-dependent hyperfine spectrum. The principles of analysis were outlined in the preceding section and entail replicate collection of ESEEM or ENDOR spectra by incremental steps and mapping the trajectory of peak positions. Deconvolution of peaks may be made either by traditional tau-suppression in the stimulated echo pulse sequence or via advanced pulse sequences such as HYSCORE (2-D ESEEM, Höfer, 1994). Mapping of spectral peak position as it varies depending on the Zeeman field is very important to the accurate determination of hyperfine terms.

ENDOR and ESEEM, under the proper circumstances, can achieve linewidths as narrow as 10 kHz, and therefore one is constrained to examining interaction energies and changes on this scale. The graphical procedure of ESEEM analysis, for which the measure of nuclear quadrupole interaction parameters is the goal, was initially described and tested in a series of papers (Flanagan & Singel, 1987; Flanagan *et al.*, 1988; Singel, 1988), and applied to a study of the Type I copper proteins (Bender *et al.*, 1997; Bender & Peisach, 1998). As a class, the Type I copper proteins are small and easily engineered by modern molecular biology methods. One member of the class, rusticyanin, was uniquely suited for a "proof of principle" study because there exists a structurally stable engineered form of the protein that lacks one of the imidazole ligands to the copper ion. Not only does this His85Ala mutant form of rusticyanin allow one to examine the ESEEM spectrum of the archetypical ^{14}N exact cancellation spectrum, but the protein binds exogenous ions and may be used to examine chemical perturbations. This section reviews the outcome of two experimental "proof of principle" studies that were made using the His85Ala rusticyanin.

The electronic structure of a molecule may be correlated to chemical reaction rates, and in this manner Hammett's concept of σ_i can, in principle, be correlated to a nuclear quadrupole coupling or hyperfine interaction constant. The analogy between ESEEM and quantum beats or level-crossing spectroscopy opens up the opportunity to measure the effective zero-field NQI parameters for quadrupolar nuclei in the reactive center (ligands or metal ion) of proteins and supramolecular assemblies. As one might expect, however, one is constrained to a system in which the energy levels can be made to cross for the probe nucleus in question, and it is therefore important to determine whether typical subtle changes in the chemical makeup of the system will yield a measurable spectroscopic change at the probe location.

Assuming that one can perform the graphical analysis, one needs to know the extent to which changes in either e_2Qq or η will shift the observed peaks in an ESEEM spectrum. Simulations indicate that the trajectories of the ZF-NQR lines are sensitive to small changes in e_2Qq and η, and the ESEEM study of the Type I copper protein stellacyanin (Bender & Peisach, 1998) demonstrated that two closely matched nuclei could be deconvoluted from the echo modulation interferogram. A peak shift of 20 kHz or more should be detected in the graphical analysis or in individual spectra (Bender et al., 1997; Bender & Peisach, 1998).

The origin of effects on NQI parameters from molecular interactions and orbital effects has been reviewed, but a representative example illustrates the problem. Nuclear quadrupole spectra are particularly attractive for the examination of internal fields because of electric field gradient (EFG) tensor eq that governs the magnitude of the coupling, and any factor that distorts the EFG will be manifest in the measured NQI. For example, motional averaging (i.e., lattice vibrations) affects NQI parameters (Duchesne, 1952), and therefore pressure and temperature affect the measured NQR spectrum. Similarly, crystal field effects may perturb the electric field gradient, and this solid-state effect was described by Bersohn (1952, 1958, 1968) and expressed mathematically by using an electrostatics model (Bertaut, 1952; Nakayama et al., 1990):

$$S = \frac{1}{4}\pi\varepsilon_0 \sum \frac{\rho(r_i)\rho(r_j)}{r_i - r_j} \tag{12}$$

This formula is conceptually simple, and its application can be facilitated, in the case of Type I copper proteins, by referral to x-ray crystallographic structures. One therefore might test this and similar models by comparing the experimentally measured NQI parameters with estimates derived from the known x-ray structures and routine computational chemistry procedures.

An NQR study of two linear polymers demonstrated that substitution of Br^- for Cl^- as the counterion can be correlated to a peak and NQI parameter shift of approximately 20 kHz (Asaji et al., 1981, 1983). A similar experiment was undertaken with the ESEEM spectrum of His85Ala rusticyanin (Bender et al., 1997), in which the normal CysMetHis2 ligand binding motif of the Type I copper center is modified by the deletion of one histidine. The still structurally and spectroscopically intact Type I protein binds ions with a correlated shift of the charge-transfer band. ESEEM spectra of protein samples His85Ala-X, where $X=H_2O$, Cl^-, Br^- revealed no shift of the effective ZF-NQR lines, and hence no solid-state effect of the type described in the preceding paragraph. But the nuclear hyperfine coupling, as measured in the double quantum peak of the ESEEM spectrum, decreased incrementally in steps of approximately 100 kHz as the substitution $H_2O \rightarrow Cl^- \rightarrow Br^-$ was made. The shift in the nuclear hyperfine interaction was attributed to increased withdrawal of electron (spin) density from the copper ion by the increasingly electronegative ligand.

A second experiment was undertaken in order to determine whether a strong DC electric field could be used to split lines in the ESEEM spectrum at exact cancellation. A linear electric field-induced hyperfine shift may be detected and understood in terms of orbital polarization. The shift has been proposed as a potential probe of the systems's excited-state wavefunction, defined as

$$\Delta\varepsilon = \sum \frac{\langle \psi_i | \mathbf{H}_E | \psi_o \rangle \langle \psi_i | \mathbf{H}_o | \psi_o \rangle}{(E_i - E_o)}$$

The electric field shift introduces characteristics of the excited-state wavefunction that, in principle, may be mapped in the same way as the spin-density distribution is assigned to orbitals of the ground-state wavefunction (Reichert, 1967; Ryde, 1976; Mims, 1976).

As a preliminary test of the hypotheses expressed in this chapter, the His85Ala rusticyanin ESEEM spectrum was recorded while a DC electric field was applied across the sample. The intent in this experiment was to create the conditions described in the preceding paragraph and thereby broaden the peaks of the usual (i.e., unperturbed) ESEEM spectrum. The three effective ZF-NQI lines of the exact cancellation ESEEM spectrum are very narrow, and one might therefore expect to see the electric field shifts that are normally measured in single-crystal or solution spectra (Reichert, 1967) despite the fact that these are powder sample spectra. The three-pulse stimulated echo ESEEM spectra are broadened as the DC electric field is increased (Figure 19), and this behavior is analogous to the electric field line broadening splitting that one observes in the case of NMR and ENDOR spectra. These data suggest that ESEEM, like ENDOR, can be used to detect small spectra line shifts that are caused by perturbations in the electric field local to a given atomic nucleus/crystallographic center.

8. CONCLUSION

Advanced EMR methods may be used to conduct quantitative measurements of nuclear hyperfine interaction energies, and these data, in turn, may be used as a tool in molecular design because of their direct relation to the frontier orbitals. The Zeeman field dependence of hyperfine spectra enables one to greatly improve the quantitative analysis of hyperfine interaction and assign numeric values to the parametric terms of the spin Hamiltonian. Graphical methods of analysis have been demonstrated that reduce the associated error that comes from a multi-parameter fit of simulations based on an assumed model. The narrow lines inherent to ENDOR and ESEEM enable precise measures of peak position and high-resolution hyperfine analyses on even powder sample materials. In particular, ESEEM can be used to obtain very narrow lines that are distributed at very nearly the zero-field NQI transition frequencies because of a quantum beating process that is associated with

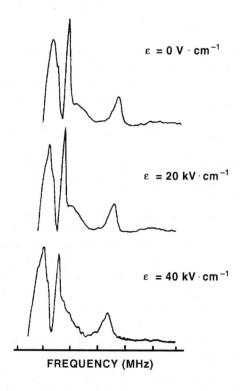

$\varepsilon = 0 \ V \cdot cm^{-1}$

$\varepsilon = 20 \ kV \cdot cm^{-1}$

$\varepsilon = 40 \ kV \cdot cm^{-1}$

FREQUENCY (MHz)

Figure 19. ^{14}N ESEEM spectra recorded near the condition of exact cancellation (i.e. effective ZF-NQI spectra) while a DC electric field was applied. Note the broadening of lines and decrease in relative peak intensity that suggest some of the contributing configurations are being shifted on account of the field.

a level-crossing process. ENDOR is equally powerful and complementary to ESEEM because of its ability to measure strong hyperfine interaction energies and derive structural information from lineshape analysis. Together, ESEEM and ENDOR are powerful tools for performing chemically chemically relevant analyses of metal complexes and free radicals.

Simulations reported in this and other reviews (Singel, 1989) demonstrate that, based on the spin-Hamiltonian model, effective zero-field nuclear quadrupole interaction parameters can be obtained by invoking a condition known as exact cancellation. On the basis of observations made concerning what experimental conditions yield optimal performance in ENDOR and ESEEM experiments, it has been suggested in this review that level crossing and the associated cross-relaxation is responsible for the deep modulation and corresponding narrow lines in the ESEEM spectrum. If level crossing and the resultant cross-relaxation processes are, in fact, the requisite condition for deep ESEEM, then the techniques described here are generally applicable. For example, any nucleus having a quadrupole moment may be subjected to ESEEM-detected ZF-NQR analysis so long as (a) the nuclear Zeeman field can be adjusted to force the levels close together, and (b) the bandwidth of the excitation pulse is sufficiently wide so that the ground states are simultaneously driven into the admixed excited state. If spin exchange can be induced between different nuclei (*i.e.,* heteronuclear cross-relaxation), it is, in principle, possible to study modulation effects of nuclei that might not be readily subjected to the aforementioned conditions.

As a prerequisite for performing the graphical analysis of ENDOR and ESEEM spectra, one needs a spectrometer that is operable over a wide frequency range. The simulations that were presented as demonstrations of the technique represent replicate spectroscopic studies that are conducted at 500-MHz increments of spectrometer operating frequency. This protocol differs from the customary multi-frequency method of recording one spectrum per octave because stepwise increments allow one to identify critical points in the hyperfine transition "trajectory" from which one may assign numeric quantities to spin-Hamiltonian parameters without relying completely on simulations. Furthermore, the flexibility of stepwise variation of operating frequency and the Zeeman energy permits one to "tune" the spectrum so that level crossing occurs, along with the associated narrowing of lines and intensity (ESEEM) enhancement. Ideally, a spectrometer should be capable of being continuously tuned over several octaves.

No commercial electron magnetic resonance spectrometer presently operates continuously over one or more frequency octaves, but instead utilizes a narrow band source that is mechanically or electrically tunable over a 500-MHz range. Despite past arguments against their use, wideband solid-state devices perform competitively with traditional narrow band devices, and numerous synthesized frequency sources are continuously tunable from 2–26 GHz. The phase noise, rated at <-80 dBc at 10 kHz offset for many systems, is excellent, and comparable to narrow band reflex klystrons. Similarly, many components can be made broadband or at least octave spanning, and therefore the experimental capability is available. One approach that is economical relies on the VXIbus instrumentation architecture, which enables one to assemble a modular instrument on a common computer-controlled mainframe that allows experimental flexibility, and further details of a broadband EMR instrument may be found in Volume 21 of this series (Bender, 2004).

Finally, quantitative analytical techniques draw their reliability from correlation to other proven methods, and quantitative EMR measures of hyperfine parameters must likewise be subjected to a form of quality assurance so that their applicability to chemical reactivity may be proven. Nuclear quadrupole interaction parameters $e_2Qq_{|zz}$ and η are recognized as powerful tools in the chemist's investigation of frontier orbital interactions. The electric field gradient, for example, reflects the hybridization, polarization, and bond order effects and will therefore vary as the covalent bonds are distorted, such as occurs when vibrational modes are excited. NQI parameters that are obtained by ESEEM or ENDOR may be refined by examining their temperature dependence (*i.e.,* the Baeyer effect, 1951; observed by

Krzytek & Kwiram, 1991) and correlated to vibrational spectra (Duchesne, 1952). Electronic distribution in molecules and classical point-charge effects may likewise be related to NQI parameters (Lucken, 1969b; Semin *et al.*, 1975), and there exists a wide body of literature for both *ab initio* and semi-empirical methods of computing NQI parameters (*cf.* Gready, 1984b). As a result, experimental measures of NQI parameters might be an excellent guide for examining the efficacy of modern and novel computational chemistry methods. One example is the QM/MM approach (Warshel & Levitt, 1976), which models large systems such as a metalloprotein by performing full *ab initio* SCF techniques to a "core" (*i.e.*, the active site) and lower accuracy molecular mechanics to the remainder of the protein. Comparison of experimental and computed NQI parameters for the various nuclei in the active site could be used to guide the design of the computational model.

9. SYMBOLS & ABBREVIATIONS

cw-	Continuous Wave
EMR	Electron Magnetic Resonance
ENDOR	Electron-Nuclear Double Resonance
ESEEM	Electron Spin Echo Envelope Modulation
NQI	Nuclear Quadrupole Interaction
NQR	Nuclear Quadrupole Resonance
ZF	Zero Field
a	(Fermi) Contact Interaction
A	Nuclear Hyperfine coupling Tensor
β	Bohr Magneton
e^2Qq_{zz}	Nuclear Quadrupole Coupling
η	Asymmetry Parameter
g	g-value (scalar quantity)
g	g-tensor
H_0	DC (primary) magnetic field (in an EMR experiment)
H_1	Magnetic (magnitude) component of the resonant microwave field
H_2	Magnetic (magnitude) component of a double resonance field (*e.g.*, rf in ENDOR)
$\mathbf{H}_{ii}$	Matrix Element (of Hamiltonian)qqq Electric field
P	Nuclear Quadrupole Coupling Tensor
τ	Interpulse spacing (ESEEM), first and second pulse
T	Interpulse spacing (ESEEM), second and third pulse
W	Energy

10. REFERENCES

Abragam A, Proctor WG. 1958. *Phys Rev* **109**:1441.

Abragam A. 1961. *The principles of nuclear magnetism*. Oxford: Oxford UP.

Adman ET. 1991. *Adv Protein Chem* **42**:145.

Alexandrov EB. 1964. *Opt Spectrosc* **17**:522.

Anderson P.W, Weiss PR. 1953. *Rev Mod Phys* **25**:269.

Ando I. 1983. *Theory of NMR parameters*. New York: Academic Press.

Antholine WE, Hanna PM, McMillan DR. 1993. *Biophys J* **64**:267.

Asaji T, Ishikawa J, Ikeda R, Inoue M, Nakamura D. 1981. *J Magn Reson* **44**:126.

Asaji T, Sakai H, Nakamura D. 1983. *Inorg Chem* **22**:202.

Baeyer H. 1951. *Z Physik* **130**:227.

Baker EN. 1988. *J Mol Biol* **203**:1071.

Balzani V, Venturi M, Credi A. 2003. *Molecular devices and machines*. Weinheim: Wiley-
    VCH.

Barfield MJ, Collins MJ, Gready JE, Sternhell S, Tansey CW. 1989. *J Am Chem Soc*
    111:4285.

Becker R. 1964. *Fields and interactions*. New York: Blaisdel.

Bender CJ. 2004. In *Biological magnetic resonance*, Vol 21. Ed CJ Bender, LJ Berliner.
    New York: Kluwer.

Bender CJ, Peisach J. 1998. *J Chem Soc, Faraday Trans* **94**:375.

Bender CJ, Sahlin M, Babcock GT, Barry BA, Chandrashekar TK, Salowe SP, Stubbe J,
    Lindström B, Petersson L, Ehrenberg A, Sjöberg B-M. 1989. *J Am Chem Soc* **111**:
    8076.

Bender CJ, Casimiro D, Peisach J, Dyson J. 1997. *J Chem Soc, Faraday Trans* **93**:3967.

Bersohn R. 1952. *J Chem Phys* **20**:1505.

Behrson R. 1958. *J Chem Phys* **29**:326.

Behrson R. 1968. *J Appl Phys* **33**:286.

Bersohn R. Shulman RG. 1966. *J Chem Phys* **45**:2298.

Bertaut EF. 1952. *J Radiat Phys* **13**:499.

Bloembergen N, Shapiro S, Persham PS, Artman JO. 1959. *Phys Rev* **114**:445.

Blum K. 1981. *Density matrix theory and applications*. New York: Plenum Press.

Böttcher R, Kirmse R, Stach J. 1984. *Chem Phys Lett* **112**:460.

Bowers KD. 1968. In *Radical ions*. Ed ET Kaiser, L Kevan. New York: John Wiley & Sons.

Bowers KD, Mims WB. 1959. *Phys Rev* **115**:285.

Bowman MK. 1992. *Israel J Chem* **32**:359

Box HC. 1977. *Radiation effects: ESR and ENDOR analysis*. New York: Academic Press.

Boys S, Bernardi F. 1970. *Mol Phys* **19**:553.

Breit G. 1933. *Rev Mod Phys* **5**:91.

Brewer RG, Hahn EL. 1973. *Phys Rev A* **8**:464.

Brown HC, Goldman G. 1962. *J Am Chem Soc* **84**:1650.

Budil D, Earle KA, Lynch B, Freed J. 1989. In *Advanced EPR: applications in biology and
    biochemistry*. Ed AJ Hoff. Amsterdam: Elsevier.

Carr HY, Purcell EM. 1954. *Phys Rev* **94**:630.

Carson RS. 1990. *RadioCommunications concepts: analog*. New York: John Wiley & Sons.

Casabella PA, Bray PJ. 1958. *J Chem Phys* **28**:1152.

Caspers WJ. 1964. *Theory of spin relaxation*. New York: Interscience.

Chebotayev VP, Dubetsky BYa. 1983. *Appl Phys B* **31**:45.

Clancy CMR, Tarasov VF, Forbes M. 1998. In *Electron paramagnetic resonance: specialist periodical reports*, Vol 16. Ed BC Gilbert, NM Atherton, MJ Davies. Cambridge: Royal Society of Chemistry.

Clark WG, Hanson ME, Lefloch F, Ségransan P. 1996. *Rev Sci Instrum* **66**:2453.

Cohen MH, Reif F. 1957. *Solid State Phys* **5**:321.

Colegrove FD, Franken PA, Lewis RR, Sands RH. 1959. *Rev Phys Lett* **3**:420.

Coremans JWA, Poluektov OG, Groenen EJJ, Canters GW, Nar H, Messerschmidt A. 1997. *J Am Chem Soc* **119**:4726.

Cornish VW, Mendel D, Schultz PG 1995. *Angew Chem Int Ed Engl* **34**:621.

Corson EM. 1951. *Perturbation methods in quantum mechanics of n-electron systems*. London: Blackie.

Cotton FA, Harris CB. 1966. *Proc Natl Acad Sci USA* **56**:12.

Dalton LR. 1985. *EPR and advanced EPR of biological systems*. Boca Raton: CRC Press.

Dalton LR, Kwiram AL. 1972. *J Chem Phys* **57**:173.

Das TP, Hahn EL. 1958. *Nuclear quadrupole resonance spectroscopy*. London: Academic Press.

Daudel R. 1973. *Quantum theory of chemical reactions*. Dordrecht: Reidel.

Daudel R, Pullman A, Salem L, Veillard A. 1982. *Quantum theory of chemical reactions*, in 3 Vols. Dordrecht: Reidel.

Davies DW. 1967. *The theory of the electric and magnetic properties of molecules*. London: John Wiley & Sons.

Deligiannakis Y, Louloudi M, Hadjiliadis N. 2000. *Coord Chem Rev* **204**:1.

Dikanov SA, Tsvetkov YuD. 1992. *Electron spin echo envelope modulation (ESEEM) spectroscopy*. Boca Raton: CRC Press.

Doyle WT, Dutton TF. 1969. *Phys Rev* **180**:424.

Duchesne J. 1952. *J Chem Phys* **20**:1804.

Dwek RA, Richards RE, Taylor D. 1969. *Ann Rev NMR Spectrosc* **2**:293.

Evans RD. 1955. *The atomic nucleus*. New York: McGraw-Hill.

Feneville S. 1977. *Rep Prog Phys* **40**:424.

Fermi E. 1930. *Z Phys* **60**:320.

Flanagan HL, Singel DJ. 1987. *J Chem Phys* **87**:5606.

Flanagan HL, Gerfen GJ, Singel DJ. 1988. *J Chem Phys* **88**:20.

Franken PA. 1961. *Phys Rev* **121**:508.

Geneaux E, Béné GJ, Perrenoud J. 1969. In *Advances in electronics and electron physics*, Vol 27. Ed L Marton. New York: Academic Press.

Gerishkin VS, Shishkin EM. 1973. In *Proceedings of the XVIIth congress ampère*. Ed V Hovi. Amsterdam: Elsevier.

Gerratt J, Cooper DL, Karadakov PB, Ramondi M. 1997. *Chem Soc Rev* **26**:87.

Geschwind S. 1967. In *Hyperfine interactions*, Ed AJ Freeman, RB Frankel. New York: Academic Press.

Gordy W. 1980. *Theory and application of electron spin resonance*. New York: John Wiley & Sons.

Gordy W, Cook RI. 1984. *Microwave molecular spectra*. New York: John Wiley & Sons.

Grampp G. 1998. In *Electron paramagnetic resonance: specialist periodical reports*, Vol 16. Ed BC Gilbert, NM Atherton, MJ Davies. Cambridge: Royal Society of Chemistry.

Grant WJC. 1964a. *Phys Rev A* **134**:1554.

Grant WJC. 1964b. *Phys Rev A* **134**:1565.

Grant WJC. 1964c. *Phys Rev A* **134**:1574.

Grant WJC. 1964d. *Phys Rev A* **135**:1265.

Gready JE. 1984a. *J Phys Chem* **88**:3497.

Gready JE. 1984b. *J Comput Chem* **5**:411.

Greiner W. 1993. *Quantum mechanics*. Berlin: Springer-Verlag.

Guibé L, Jugie G. 1981. In *Molecular interactions*, Vol 2. Ed H Ratajczak, WJ Orville-Thomas. Chichester: John Wiley & Sons.

Gutkowsky HS, McCall DW, McGarvey BR, Meyer LH. 1952. *J Am Chem Soc* **74**:4809.

Gutkowsky HS, Woesner DE. 1958. *Phys Rev Lett* **1**:6.

Hahn EL. 1950. *Phys Rev* **80**:580.

Hahn EL, Maxwell DR. 1951. *Phys Rev* **84**:1246.

Hamilton AD, Ed. 1996. *Supramolecular control of structure and reactivity*. Chichester: John Wiley & Sons.

Hammett LP. 1970. *Physical organic chemistry*. New York: McGraw-Hill.

Harriman JE. 1978. *Theoretical foundations of electron spin resonance*. New York: Academic Press.

Höfer P, Grupp A, Mehring M. 1986. *Phys Rev A* **33**:3519.

Höfer P. 1994. *J Magn Reson A* **111**:77.

Hoffman BM, Martinson J, Venters JA. 1984. *J Magn Reson* **59**:110.

Hoffman BM, DeRose VJ, Doan PE, Gurbiel RJ, Houseman ALP, Telsev J. 1993. *Biol Magn Reson* **13**:151.

Hollas JM. 1982. *High resolution spectroscopy*. London: Butterworths.

Hurst GC, Henderson TA, Kreilick RW. 1985. *J Am Chem Soc* **107**:7294.

Hyde JS. 1967. In *Magnetic resonance in biological systems*, Ed A Ehrenberg, BG Malström, T Vänngård. Oxford: Pergamon Press.

Hyde JS, Froncisz W. 1982. *Ann Rev Biophys Bioeng* **11**:391.

Hyde JS, Rist GH, Eriksson IR. 1968. *J Chem Phys* **72**:4269.

Iwasaki M, Toriyama K, Nunome K. 1986. *J Chem Phys* **86**:5971.

Jaffé HH. 1952a. *J Chem Phys* **20**:279.

Jaffé HH. 1952b. *J Chem Phys* **20**:778.

Jaffé HH. 1952c, *J Chem Phys* **20**:1554.

Jarrett HS. 1956. *J Chem Phys* **25**:1289.

Jiang X-K. 1997. *Acc Chem Res* **30**:283.

Kaupp M, Bü M, Malkin VG. 2004. *Calculation of nmr and epr parameters: theory and applications*. Weinheim: Wiley-VCH.

Kauzmann W. 1957. *Quantum chemistry*. New York: Academic Press.

Kelly RL. 1966. *Phys Rev* **147**:376.

Kevan L, Kispert LD. 1976. *Electron spin double resonance spectroscopy*. New York: John Wiley & Sons.

Kevan L, Schwartz RN. 1979. *Time domain electron spin resonance*. New York: Wiley.

Kitaura K, Morokuma K. 1976. *Int J Quantum Chem* **10**:325.

Klauder JR, Anderson PW. 1962. *Phys Rev* **125**:912.

Klauder JR, Sudarshan ECG. 1968. *Fundamentals of quantum optics*. New York: W.A. Benjamin.

Klauder JR, Skagerstam B-S. 1985. *Coherent states: applications in physics and mathematical physics*. Singapore: World Scientific.

Kollman PA. 1978. In *Modern theoretical chemistry*, Vol 4. Ed HF Schaefer. New York: Plenum Press.

Kopfermann H. 1958. *Nuclear moments*. New York: Academic Press.

Kroto HW. 1975. *Molecular rotation spectra*. London: John Wiley & Sons.

Krzytek J, Kwiram AL. 1991. *J Am Chem Soc* **113**:9768.

Krzytek J, Notter M, Kwiram AL. 1994. *J Phys Chem* **98**:3559.

Krzytek J, Kwiram AB, Kwiram AL. 1995. *J Phys Chem* **99**:402.

Levanon H, Möbius K. 1997. In *Annual review of biophysics and biomolecular structure*, Vol 26. Ed RM Stroud, WL Hubbell, WK Olson, MP Sheetz. Palo Alto: Annual Reviews Inc.

Lucken EAC. 1969a. *Struct Bonding* **6**:1.

Lucken EAC. 1969b. *Nuclear quadrupole coupling constants*. London: Academic Press.

Mabbs FE, Collison D. 1992. *Electron paramagnetic resonance of d-transition metal ions*. Amsterdam: Elsevier.

Macomber JD. 1976. *The dynamics of spectroscopic transitions*. New York: John Wiley & Sons.

Mataga N. 1981. In *Molecular interactions*, Vol 2. Ed H Ratajczak, WJ Orville-Thomas. Chichester: John Wiley & Sons.

Mateescu GhD. 1993. *2D NMR: density matrix and product operator treatment*. Engelwood Cliffs: Prentice-Hall.

McConnell HM. 1956. *J Chem Phys* **24**:764.

McConnell HM, Strathdee J. 1959. *Mol Phys* **2**:129.

McConnell HM, Weaver HE. 1956. *J Chem Phys* **25**:307.

Meal HC. 1952. *J Am Chem Soc* **74**:6121.

Mehring M, Höfer P, Grupp A. 1986. *Phys Rev A* **33**:3523.

Memory JD. 1968. *Quantum theory of magnetic resonance parameters*. New York: McGraw-Hill.

Mercks RPJ, DeBeer R. 1979. *J Chem Phys* **83**:3319.

Michelson AA. 1903. *Light waves and their uses*. Chicago: U Chicago P.

Mims WB. 1968. *Phys Rev* **168**:370.

Mims WB. 1972a. In *Electron paramagnetic resonance*, Ed S Geschwind. New York: Plenum Press.

Mims WB. 1972b. *Phys Rev B* **5**:2409.

Mims WB. 1972c. *Phys Rev B* **6**:3543.

Mims WB. 1976. *The linear electric field effect in paramagnetic resonance*. Oxford: Clarendon Press.

Mims WB. 1984. *J Magn Reson* **59**:291.

Mims WB, Peisach J. 1978. *J Chem Phys* **69**:4921.

Mims WB, Peisach J, Davies JL. 1971. *J Chem Phys* **66**:5536.

Mims WB, Davis JL, Peisach J. 1984. *Biophys J* **45**:755.

Morokuma K. 1971. *J Chem Phys* **55**:1236.

Morokuma K, Kitaura K. 1980. In *Molecular interactions*, Vol 1. Ed H Ratajczak, WJ Orville-Thomas. Chichester: John Wiley & Sons.

Murai H, Tero-Kubota S, Yamaguchi S. 2000. In *Electron paramagnetic resonance: specialist periodical reports*, Vol 17. Ed BC Gilbert, MJ Davies, KA McLauchlan. Cambridge: Roayl Society of Chemistry.

Nakayama H, Saito K, Kishita M. 1990. *Z Naturforsch* **45a**:275.

Nakatsuji H. 1993. In *Nuclear magnetic shielding and molecular structure*, Ed JA Tossell. Dordrecht: Kluwer.

Nakatsuji H, Kanda K, Endo K, Yonezawa T. 1984. *J Am Chem Soc* **106**:4653.

Nakatsuji H, Nakao T, Kanda K. 1987. *Chem Phys* **118**:25.

Norris JR, Thurnauer MC, Bowman MK. 1980. *Adv Biol Med Phys* **17**:365.

Nyquist H. 1946. *AIEE Trans* 617.

O'Malley PJ, Babcock GT. 1980. *J Chem Phys* **80**:3912.

Ponti A, Schweiger A. 1994. *Appl Magn Reson* **7**:363.

Poole CP, Farach HA. 1971. *Relaxation in magnetic resonance: dielectric, and Mössbauer applications*. New York: Academic Press.

Poole CP, Farach HA. 1987. *Theory of magnetic resonance*. New York: John Wiley & Sons.

Prisner TF. 1997. *Adv Magn Opt Res* **20**:245.

Ramsey NF. 1950. *Phys Rev* **78**:699.

Ratajczak H, Orville-Thomas WJ, Eds. 1982. *Molecular interactions*, in 3 Vols. Chichester: John Wiley & Sons.

Rédei LB. 1963. *Phys Rev* **130**:420.

Reichert JF. 1967. In *Hyperfine interactions*, Ed AJ Freeman, RB Franken. New York: Academic Press.

Rist GH, Ammeter J, Günthard HsH. 1968. *J Chem Phys* **49**:2210.

Rist GH, Hyde JS. 1970. *J Chem Phys* **52**:4633.

Rowan LG, Hahn EL, Mims WB. 1965. *Phys Rev A* **137**:61.

Ryde N. 1976. *Atoms and molecules in electric fields*. Stockholm: Almqvist & Wiksell.

Sauvage J-P. 2001. *Molecular machines and motors*. Berlin: Springer-Verlag.

Schempp E, Bray PJ. 1970. In *Physical chemistry: an advanced treatise*, Vol 4. Ed H Eyring, D Henderson, W Jost. New York: Academic Press.

Schweiger A, Rudin M, Günthard HsH. 1979. *Mol Phys* **37**:1573.

Schweiger A. 1982. *Struct Bonding* **51**:1.

Scrocco E, Thomasi J. 1964. In *Molecular orbitals in chemistry, physics, and biology*. Ed P-O Löwdin, B Pullman. New York: Academic Press.

Semin GK, Bubushkina TA, Yakobson GG. 1975. *Nuclear quadrupole resonance in chemistry*. New York: John Wiley & Sons.

Schuster P, Zundel G, Sandorfy C. 1976. *The hydrogen bond*, in 3 Vols. Amsterdam: North-Holland.

Segal GA, Ed. 1977. *Semiempirical methods of electronic structure calculation, Modern Theoretical chemistry*, Vols 7, 8. New York: Plenum Press.

Shubin AA, Dikanov SA. 1983. *J Magn Reson* **52**:1.

Sinanoglu O, Ed. 1965. *Modern quantum chemistry*. New York: Academic Press.

Singel DJ. 1989. In *Advanced EPR: applications in biology and biochemistry* Ed AJ Hoff. Amsterdam: Elsevier.

Slater JC. 1968. *Quantum theory of matter.* New York: McGraw-Hill.

Squire PT. 1965. *Proc Phys Soc* **86**:573.

Standley KJ, Vaughan RA. 1966. *Electron spin relaxation phenomena in solids.* New York: Harper & Row.

Tanford C, Reynolds J. 2001. *Nature's robots.* New York: Oxford UP.

Thomann H, Dalton LR, Dalton L. 1984. In *Biological magnetic resonance,* Vol 6. Ed LJ Berliner, J Ruebens. New York: Plenum Press.

Tidwell TT, Rappaport Z, Perrin CL, Eds. 1997. *Physical organic chemistry for the 21st century: a symposium in print. Pure and Appl Chem* **89**:211.

Townes CH, Dailey BP. 1955. *J Chem Phys* **23**:118.

Tsvetkov YuD, Dikanov SA. 1987. In *Metal ions in biological systems*, Vol 22. Ed H Sigel. New York: Dekker.

Un S, Brunel LC, Brill TM, Zimmerman J-L, Rutherford AW. 1994. *Proc Natl Acad Sci USA* **91**:5262.

van de Kamp M, Canters GW, Andrew CR, Sanders-Loehr J, Bender CJ, Peisach J. 1993. *Eur J Biochem* **218**:229.

van Vleck JH. 1952. *Ann NY Acad Sci* **55**:928.

Verstelle JC. 1968. In *Handbuch der physik*, Band 18.1. Ed HPJ Wijn. Berlin: Springer-Verlag.

Viehe HG, Janousek Z, Merényi R, Eds. 1986. *Substituent effects in radical chemistry.* NATO ASI C 189. Boston: Kluwer.

Warncke K, Brooks HB, Lee HI, McCracken J, Davidson VL, Babcock GT. 1995. *J Am Chem Soc* **117**:10063.

Warshel A, Levitt M. 1976. *J Mol Biol* **103**:227.

Webb GA, Witanowski M. 1985. In *Topics in molecular interactions*. Ed WJ Orville-Thomas, H Ratajczak, CNR Rao. Amsterdam: Elsevier.

Weiss A. 1988. In *Magnetic resonance and related phenomena. Proceedings of the xxiv congress ampère.* Ed J Stankowski, N Piâslewski, S Idziak. Amsterdam: Elsevier.

Weiss A, Wigand S. 1990. *Z Naturforsch* **45a**:195.

Weissman SI. 1956. *J Chem Phys* **25**:890.

Weltner W. 1983. *Magnetic atoms and molecules.* New York: Scientific and Academic Editions.

Wilson EB. 1952. *Ann NY Acad Sci* **55**:943.

CHAPTER 4

NEW METHODS OF SIMULATION OF Mn(II) EPR SPECTRA: SINGLE CRYSTALS, POLYCRYSTALLINE AND AMORPHOUS (BIOLOGICAL) MATERIALS

Sushil K. Misra

Physics Department, Concordia University, 1455 de Maisonneuve Boulevard West, Montreal, Québec H3G 1M8, Canada

1. INTRODUCTION

Biological systems exhibit properties of amorphous materials. The Mn(II) ion in amorphous materials is characterized by distributions of spin-Hamiltonian parameters around mean values. It has a certain advantage over other ions, being one of the most abundant elements on the earth. The extent to which living organisms utilize manganese varies from one organism to the other. There is a fairly high concentration of the Mn(II) ion in green plants, which use it in the O_2 evolution reaction of photosynthesis (Sauer, 1980). Structure-reactivity relationships in Mn(II)–O_2 complexes are given in a review article by Coleman and Taylor (1980). Manganese is a trace requirement in animal nutrition; highly elevated levels of manganese in the diet can be toxic, probably because of an interference with iron homeostasis (Underwood, 1971). On the other hand, animals raised with a dietary deficiency of manganese exhibit severe abnormalities in connective tissue; these problems have been attributed to the obligatory role of Mn(II) in mucopolysaccharide metabolism (Leach, 1971). Mn(II) has been detected unequivocally in living organisms. Examples are Mn(II)-metalloproteins, such as pyruvate carboxylase (Sutton *et al.*, 1966), plant lectins (Galbraith & Goldstein, 1970), dioxygenase enzyme isolated from *Bacillus brevis* (Que *et al.*, 1981), arginase isolated from liver tissue (Hirsch-Kolb *et al.*, 1971), Mn(II) protein glutamine synthetase isolated from sheep brain (Wedler *et al.*, 1982), placental diamine oxidase (Crabbe *et al.*, 1976), clam shells and sea shells (Blanchard & Chasteen, 1976; White *et al.*, 1982). (See, among others, the review article by McEuen (1982), which summarizes information on Mn-metalloproteins and several Mn-activated enzymes.) The free Mn(II) level in rat livers is higher in fed than in fasted animals, with the total amount of Mn(II) being very nearly constant (Ash &

Mn(II) being very nearly constant (Ash & Schramm, 1982). The ability of Mn(II) to substitute for Mg(II) in a wide variety of enzyme reactions has made Mn(II) popular as a spectroscopic probe in many enzyme complexes. There are a number of similarities in the coordination properties of Mn(II) and Mg (II) ions; in many cases the maximal velocities of enzyme reactions activated by Mn(II) are nearly equivalent to those obtained with Mg(II). Mn(II), thus, remains one of the best surrogates for Mg(II) in studies of enzymic complexes.

Mn(II) EPR spectra in biological systems are very much like those in glasses — e.g., that in lithium-borate glass (Griscom & Griscom, 1967) matches closely that in kinase oxalate ternary complex (Reed & Markham, 1984; referred to hereafter as RM, and references therein). Table 1 lists the measured values of the spin-Hamiltonian parameters parameters (g, D, E) in some proteins, as taken from RM.

Table 1. Mn(II) spin-Hamiltonian (SH) parameters in metalloproteins after Reed and Markham (1984) (references given therein), indicated by RM. Additional references are as indicated. The D, E, A values are in units of Gauss (as calculated by dividing the value in energy units by $g_e \mu_B$, unless otherwise indicated; $\eta = D/E$). The spectra were taken at the X-band, unless otherwise stated.

Host	SH Parameter	Remarks
Concanavalin A	$g = 2.0009$ $D = 232$ $\eta = 0.185$ $A_\parallel = 94.4$ $A_\perp = 91.5$	Single crystal (RM)
Concanavalin A	$g = 2.0007$ $D = 230$ $\eta = 0.11$	Solution (RM)
Creatine kinase, ADP, creatine, and formate complex	$D = 300$; $\eta = 0.06$	A Gaussian spread of 12 G was used in simulation. Matrix-diagonalization simulation with $D = 315$ G led to a better fit (Coffino & Peisach, 1996)
Glutamine synthetase (GS) with Mn(II) and methionine sulfoximine (MSOX)	$D = 140$	D estimated from the relative intensity of allowed and forbidden transitions of the central sextet (RM)
GS-Mn(II)-MSOX-MgADP complex	$D \sim 150$	
Hadacidin (N-formyl hydroxy amino acetate)-bound ternary enzyme-Mn(II)-GTP (or -GDP) complex in the presence of IMP	$D = 1,000$	Presence of IMP increases D significantly (9.35 GHz) (RM)

Mn(II)-α-lactalbumin complex	$D < 0.02$ cm^{-1}	9 GHz (frozen, 77 K), 35 GHz (283 K). The spectrum did not narrow when the temperature was increased to ambient temperature. Spectra remarkably similar to those for Mn(II) -troponin or -parvalbumin complexes. Spectra indicate a relatively highly symmetric (cubic) environment around the Mn(II) ion. Fine structure unresolved due to relatively small value of D (Berliner *et al*, 1993).
Mn(II), pyruvate kinase, and phospho enolpyruvate (ternary complex)	$D = 1300$; $|\eta| = 1/3$	35 GHz; Closely matches Mn(II) spectrum in lithium borate glasses. (Weak interaction of Mn(II) with C-O-P bridge oxygen of P-enolpyruvate upsets the electronic symmetry around the metal ion leading to large values of D, E.) (RM)
Nitrate-bound enzyme-Mn(II)-GDP-IMP complex	$D = 360$ $E = 120$	35 GHz. Proved fine-structure splittings due to nitrate's role as a transition state analog for the phosphoryl transfer step of the reaction. (RM)
Ternary enzyme-Mn(II)-GTP with co-substrate aspartate	$D = 260$ $E = 60$	Spectra are best fitted with a distribution in D with a half width of 20 G. (Same values of parameters in the enzyme with GDP complex (RM).)

Being an S-state ion, the Mn(II) g and A matrices can be considered to be isotropic in so far as simulation of spectra in amorphous materials is concerned, as their small anisotropies are smeared out in these samples. On the other hand, the hyperfine structure of its spectrum adds some complexity to the spectrum, which also reflects the effect of interaction with the environment, since the Mn(II) ion possesses an electronic spin larger than ½.

Simulation of polycrystalline (or powder; these terms are used interchangeably) spectra of the Mn(II) ion has been of great interest recently, especially in metalloproteins (references on metalloproteins include Reed & Markham, 1984; Chiswell *et al.*, 1987; for EPR crystallography of metalloproteins see Chien & Dickinson, 1981; for ESR of iron proteins, Smith & Pilbrow, 1980; for X- and Q-band EPR studies of Cu^{2+}, VO^{2+}, and Gd^{3+} ions in bovine α-lactalbumin complexes, see Musci, Reed, and Berliner, 1986; for copper proteins see Boas, 1984; Beinert, 1985; Palmer, 1985; Hanson & Pilbrow, 1987; Blumberg & Peisach, 1987; Hanson & Wilson, 1988; Villafranca & Raushel, 1990). Mn(II) EPR spectra in amorphous materials can be simulated by superposition of single-crystal EPR spectra obtained for various orientations of the external Zeeman field over the unit sphere (see Misra & Sun, 1991, for a review of published single-crystal Mn(II) EPR results). To simulate the Mn(II) spectrum in an amorphous material, one takes into account distributions of zero-field splitting parameters D and E, as well as random orientations of the Mn(II) principal magnetic axes with respect to the external magnetic field.

There are two approaches to simulate and interpret Mn(II) EPR spectra in amorphous materials. A simple approach is to take into account second- or third-order perturbation expressions for eigenvalues and relative intensities of EPR lines for different site symmetries, *e.g.*, axial (Misra, 1994) and orthorhombic (Misra,

1997). Information on the values of the parameters (D, E) can then be obtained from the "peaks" of the spectra. A more rigorous approach is to simulate the spectrum on a computer, using appropriate line shapes and distribution of parameters, (see, *e.g.,* Kliava and Purans, 1980; Misra, 1996). The purpose of this article is to provide details of how to simulate EPR spectra in amorphous materials, with the objective to obtain knowledge of Mn(II) zero-field splitting (ZFS) parameters in these materials. Section 2 deals briefly with the details of how to simulate Mn(II) EPR spectra in a polycrystalline material, assuming sharp values for parameters D and E, based on expressions up to third order in perturbation as well as those obtained by diagonalization of the spin-Hamiltonian matrix on a computer. This is then exploited in §3 to simulate spectra in amorphous materials, *e.g.,* glasses, taking into account distributions of D, E values. Computer-simulated spectra and their comparison with experimental spectra in glasses, as prototypes for biological systems, are described in §4, while concluding remarks are offered in §7.

2. SINGLE-CRYSTAL AND POLYCRYSTALLINE Mn(II) SPECTRA

2.1. Single-Crystal Spin-Hamiltonian and Its Eigenvalues for Orthorhombic Distortion

This section presents expressions for the energies of the Mn(II) ion in a single crystal, calculated up to third order in perturbation, along with allowed hyperfine line positions and hyperfine forbidden hyperfine-doublet (FHD) separations. These are then exploited to calculate FHD separations, and to estimate D and E parameters from a polycrystalline EPR spectrum depending on FHD and peaks of allowed h.f. lines in the central sextet. The various possible shapes of Mn(II) EPR spectrum in polycrystalline and glass samples are shown in Figure 1. It is noted that only the lines belonging to the central hyperfine sextet are usually observed in polycrystalline/amorphous materials, and that often the allowed and/or forbidden lines may or may not be resolved.

The spin-Hamiltonian for the Mn(II) ion (electronic spin $S = 5/2$, nuclear spin $I = 5/2$) in a quadratic crystal field for orthorhombic distortion, neglecting terms of fourth order in electronic spin operator, is given by

$$H = g\mu_B \mathbf{B} \cdot \mathbf{S} - g_N \mu_N \mathbf{B} \cdot \mathbf{I} + A\mathbf{I} \times \mathbf{S} + H_{ZFS} \tag{1}$$

where the first three terms represent the electronic-Zeeman, nuclear-Zeeman, and electron-nuclear hyperfine interactions, respectively. Here μ_B and μ_N are Bohr and nuclear magnetons, respectively; and $\mathbf{B}$ is the external (Zeeman) magnetic field; H_{ZFS} is the zero-field splitting (ZFS) term, which can be written for non-axial situation as

$$H_{ZFS} = D\left\{ S_z^2 - \frac{1}{3}S(S+1) \right\} + \frac{E}{2}\left\{ S_+^2 + S_-^2 \right\} \tag{2}$$

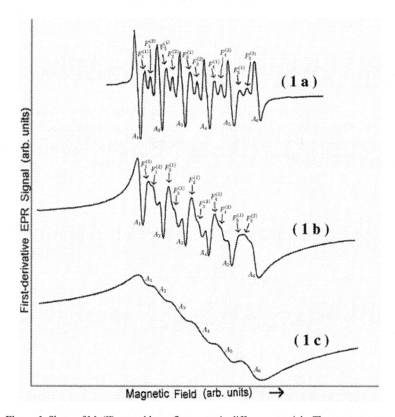

Figure 1. Shape of Mn(II) central hyperfine sextet in different materials. The spectra are exhibited in (a–c), respectively, in a single crystal, polycrystalline sample, and disordered material only to demonstrate their qualitative features. It is noted that while the line positions correspond to each other in (b) and (c) as shown, they do not exactly correspond to those in (a). (a) A typical X-band (~9.5 GHz) Mn(II) central hyperfine sextet in a single crystal. The allowed hyperfine lines — A1, A2, A3, A4, A5, and A6 — as well as the five hyperfine forbidden doublets — $F_1^{(1)}$, $F_1^{(2)}$, $F_2^{(1)}$, $F_2^{(2)}$, $F_3^{(1)}$, $F_3^{(2)}$, $F_4^{(1)}$, $F_4^{(2)}$, $F_5^{(1)}$, and $F_{51}^{(2)}$ — in order of increasing value of the magnetic field at which they occur, have been indicated. For identification of the transitions corresponding to these lines, see the text in §3. (b) A typical X-band Mn(II) central hyperfine sextet in a polycrystalline sample. The various allowed and hyperfine lines corresponding to those in a single crystal as described in the caption of (a) have been indicated. (c) A typical X-band Mn(II) central hyperfine sextet for a disordered material, wherein the value of D varies over different parts of the sample, resulting in complete smearing of the hyperfine forbidden lines and a significant broadening of the hyperfine allowed lines. Adapted with permission from Misra (1994).

where z denotes the principal axis of the second-order zero-field splitting (ZFS) tensor; D and E are the axial and orthorhombic ZFS parameters; and $S_\pm = S_x \pm iS_y$. Matrices $\tilde{g}$ and $\tilde{A}$ for the S-state Mn(II) ion have here been assumed isotropic. The spin-Hamiltonian can now be written as the sum of zero-order and perturbation parts:

$$H = H^{(0)} + H^{'} \tag{3}$$

with

$$H^{(0)} = g\mu_B B S_z - g_N \mu_N \mathbf{B} \cdot \mathbf{I}, \text{ and } H^{'} = H_{ZFS} + \mathbf{AI} \cdot \mathbf{S} \tag{4}$$

where the z-axis, assumed parallel to the direction of the external (Zeeman) magnetic field, $\mathbf{B}$, has been chosen as the axis of quantization. Expressions for the perturbation energies, up to third order in perturbation, are (Meirovitch & Popko, 1978; Markham et $al.$, 1979)

$$E_{Mm} = E_{Mm}^{(0)} + E_{Mm}^{(1)} + E_{Mm}^{(2)} + E_{Mm}^{(3)}$$

with

$$E_{Mm}^{(0)} = g\mu_B B M; \tag{5}$$

$$E_{Mm}^{(1)} = AMm - g_N\mu_N Bm + a\left\{ M^2 - \frac{1}{3}S(S+1) \right\}; \tag{6}$$

$$E_{Mm}^{(2)} = \frac{A^2}{2G}\left\{ MI(I+1) - mS(S+1) + M^2m - Mm^2 \right\} +$$
$$\frac{2b_+b_-}{G}\left\{ 8M^3 + M - 4MS(S+1) \right\} - \tag{7}$$
$$\frac{2c_+c_-}{G}\left\{ 2M^3 + M - 2MS(S+1) \right\};$$

$$E_{Mm}^{(3)} = \frac{Ab_+b_-}{G^2}\left\{ S_M^+(2M+1)^3 - S_M^-(2M-1)^3 \right\} +$$
$$\frac{Ac_+c_-}{G^2}\left\{ (M+1)S_M^+S_{M+1}^+ - (M-1)S_M^-S_{M-1}^- \right\} +$$
$$\frac{Re(b_+^2c_-)}{G^2}\left\{ (2M+1)(2M+3)S_M^+S_{M+1}^+ + \right.$$
$$\left. (2M-1)(2M-3)S_M^-S_{M-1}^- - 2(2M-1)(2M+1)S_M^-S_M^+ \right\} +$$
$$\frac{A^3}{4G^2}\left\{ (-M+m-1)S_M^+I_m^- + (M-m-1)S_M^-I_m^+ \right\} + \tag{8}$$
$$\frac{Ab_+b_-}{G^2}\frac{2m}{M}\left\{ \left[S(S+1) - M^2 \right]^2 - M^2 \right\} +$$
$$\frac{Ac_+c_-}{2G^2}m\left\{ S_M^+S_{M+1}^+ - S_M^-S_{M-1}^- \right\} +$$
$$\frac{A^2a}{4G^2}\left\{ (2M+1)S_M^+I_m^- - (2M-1)S_M^-I_m^+ \right\}$$

where

$$a = D\left\{\left(\frac{3\cos^2\theta-1}{2}\right) + \frac{3}{2}\eta\sin^2\theta\cos 2\phi\right\}$$

$$b_{\pm} = \frac{D}{4}\left\{-\sin 2\theta + \eta\sin 2\theta\cos 2\phi \pm i2\eta\sin\theta\sin 2\phi\right\}$$

$$c_{\pm} = \frac{D}{4}\left\{\sin^2\theta + \eta(\cos^2\theta+1)\cos 2\phi \pm i2\eta\cos\theta\sin 2\phi\right\}$$

$$\eta = \frac{E}{D}$$

$$G = g\mu_B B$$

$$S_M^{\pm} = S(S+1) - M(M\pm 1)$$

$$I_m^{\pm} = I(I+1) - m(m\pm 1) \tag{9}$$

In the above equations, the order of perturbation is indicated by number n ($= 0$, 1, 2, 3) within brackets in the superscripts on E_{Mm}; M and m are the electronic and nuclear magnetic quantum numbers, respectively; θ and ϕ are, respectively, the polar and azimuthal angles of the principal axis of the ZFS tensor with respect to the external magnetic field; and Re denotes the real part. Because of its small magnitude, only the first-order term of the nuclear Zeeman energy is taken into account. The range for η is $0 \leq \eta \leq 1/3$.

The positions of the allowed — ($1/2$, $m \leftrightarrow -1/2$, m) — and forbidden, M, m $\leftrightarrow M$, $m-1$ ($\Delta M=0$, $\Delta m=-1$), or M, $m-1 \leftrightarrow M$, m ($\Delta M=0$, $\Delta m=1$), hyperfine transitions belonging to the Mn(II) central sextet can be expressed from eqs. (5)–(8) by the use of resonance condition $h\eta = E_{1/2,m} - E_{-1/2,m} = G_0 = g\mu_B B_0$, where B_0 is the magnetic field corresponding to the middle of the central sextet, as follows:

$$B_c(m,\theta,\phi) = \left[1 - \frac{A}{G_o}m - \frac{4A^2m}{G_o^3}Df_1(\theta,\phi)\right.$$

$$+ \frac{D^2}{G_o^3}\left\{(4G_o - 36Am)f_3(\theta,\phi) - 2(G_o - Am)f_2(\theta,\phi)\right\}$$

$$\left. - \frac{A^2}{8G_o^3}\left\{(35-4m^2)G_o + m(4m^2-65)A\right\}\right]B_o \tag{10}$$

In eq. (13) the expressions for $f_1(\theta,\phi)$, $f_2(\theta,\phi)$, and $f_3(\theta,\phi)$ are given as follows:

$$f_1(\theta,\phi) = \left\{(3\cos^2\theta-1)+3\eta\sin^2\theta\cos 2\phi\right\} \tag{11}$$

$$f_2(\theta,\phi) = \left\{[\sin^2\theta+\eta\cos 2\phi(1+\cos^2\theta)]^2 + 4\eta^2\cos^2\theta\sin^2 2\phi\right\} \tag{12}$$

$$f_3(\theta,\phi) = \left\{(1-\eta\cos 2\phi)^2\sin^2 2\theta + 4\eta^2\sin^2\theta\sin 2\phi\right\} \tag{13}$$

2.2. Forbidden HFD Separations

The positions of the forbidden hyperfine lines $(1/2, m \leftrightarrow -1/2, m)$ denoted by $B_+(M,m)$ and $B_-(M,m)$ in the Mn(II) central sextet are defined as follows, separating the contributions into various orders of perturbation:

$$B_+(M,m) = B(M, m \leftrightarrow M-1, m-1) \tag{14}$$

$$B_-(M,m) = B(M, m-1 \leftrightarrow M-1, m) \tag{15}$$

Finally, using eqs. (14) and (15), and for simplification replacing $B_+(M,m)$ and $B_-(M,m)$ by B_0 in the denominators, the five hyperfine forbidden doublet separations turn out to be (for $m = -3/2, -1/2, 1/2, 3/2, 5/2$) (Misra, 1994, 1997)

$$\Delta B(M,m) = B_+(M,m) - B_-(M,m)$$
$$= \Delta B^{(1)}(M,m) + \Delta B^{(2)}(M,m) + \Delta B^{(3)}(M,m) \tag{16}$$

where

$$\Delta B^{(1)}(M,m) = -\frac{A(2M-1)}{g\mu_B} \tag{17}$$

$$\Delta B^{(2)}(M,m) = \frac{17A^2}{2g^2\mu_B^2 B_0(M)}, \tag{18}$$

$$\Delta B^{(3)}(M,m) = \frac{D^2 A f_2(\theta,\varphi)}{g^3 \mu_B^3 B_0(M)^2} + \frac{4A^2 D}{g^3 \mu_B^3 B_0(M)^2} f_1(\theta,\varphi)(2m-1)$$
$$- \frac{67A^3}{4g^3 \mu_B^3 B_0(M)^2}(2m-1). \tag{19}$$

In eqs. (18) and (19), the expressions for $f_1(\theta,\varphi)$ and $f_2(\theta,\varphi)$ are the same as those given by eqs. (11) and (12), respectively. The superscripts in (17)–(19) represent the order of the contribution in perturbation to the eigenvalues of the spin-Hamiltonian. Further, it is noted that for the central sextet ($M = 1/2$), $\Delta B^{(1)}(M,m)$, is zero, while $\Delta B^{(2)}(M,m)$ is positive regardless of the absolute sign of the hyperfine-interaction constant (A). This means that the forbidden line position, $B_+(M,m)$, is at a higher magnetic field value than the forbidden line position, $B_-(M,m)$, unless the value of D is sufficiently large to render $\Delta B^{(3)}$ the dominant contribution.

2.3. Estimation of Parameters D and E from a Polycrystalline Spectrum

FHD Separations. The expression for the forbidden hyperfine line positions, $B_c(m,\theta,\varphi)$, listed in §2.2 (i) above, are those for a single crystal, whose axis of symmetry is oriented at angle θ with respect to the external magnetic field (**B**). In a polycrystalline material, the constituting crystallites are randomly oriented, so that all values of θ from 0 to 90° are possible, and the number of crystallites with their axes oriented between θ and $\theta + d\theta$ is proportional to $\sin\theta \, d\theta$. Similarly, all values of ϕ between 0 and 2π are possible, and the number of crystallites with their axes oriented between ϕ and $\phi + d\phi$ is proportional to $d\phi$. The doublet separation corresponding to the peaks (maxima) of forbidden hyperfine lines (FHDs) for a polycrystalline sample are calculated at angle $\theta = (0$, at which the maximum value of f0((,() sin (occurs as a function of (, since the intensity of a forbidden hyperfine line M, m symbol 171 \f "Symbol" \s 12 « M, m − 1 ((M = 0, (m = −1), or M, m − 1 symbol 171 \f "Symbol" \s 12 « M, m ((M = 0, (m = 1), relative to the allowed hyperfine line $M, m \leftrightarrow M - 1, m$ ($\Delta M = 1, \Delta m = 0$) in the central sextet ($M = 1/2$) is given by the following expression:

$$R(\theta,\varphi) = \left(\frac{3D}{4g\mu_B B_0(M)}\right)^2 f_0(\theta,\varphi)\left\{1+\left[\frac{S(S+1)}{3M(M-1)}\right]\right\}^2 \times \quad (20)$$

$$[I(I+1) - m(m-1)],$$

where $f_0(\theta,\varphi) = \sin^2 2\theta \, (1 - \eta \cos\varphi)^2$.

A simple calculation yields the values of the angular-dependent factors in the forbidden line positions, given by eqs. (14) and (15), corresponding to the maximum of intensity, which occurs, respectively, for $\theta = \theta_0$, i.e. when

$$\cos^2\theta_o = \frac{4(1-\eta^2)}{5(2+\eta^2)} \quad (21)$$

and

$$\sin^2\theta_0 = \frac{3(2+3\eta^2)}{2(2+\eta^2)} \quad (22)$$

Finally, the separation of peaks (maxima of intensity) corresponding to doublets of hyperfine forbidden transitions) in the central sextet ($M = 1/2 \leftrightarrow M = -1/2$) in a polycrystalline sample turns out to be from Equations (16) - (19):

$$\Delta B^{(PC)}\left(\frac{1}{2},m\right) = \left[\frac{17A^2}{2G_0^2} + \frac{9D^2 AF(\eta)}{25G_0^3} + \frac{4A^2 D}{5G_0^3}(2m-1) - \frac{67A^3}{4G_0^3}(2m-1)\right]B_o \quad (23)$$

with

$$F(\eta) = \left\{\left(\frac{2+3\eta^2}{2+\eta^2}\right)^2 + \eta^2\left[\frac{25}{18} + \frac{10(2-\eta^2)}{3(2+\eta^2)} + \frac{4}{9}\left(\frac{2-\eta^2}{2+\eta^2}\right)^2\right]\right\} \quad (24)$$

Equation (23) is the key equation to be used to estimate the values of D and E from Mn(II) EPR spectra in an amorphous material. Figure 2 shows plots of the values of $|D|$ as functions of average hyperfine forbidden-doublet separation in the

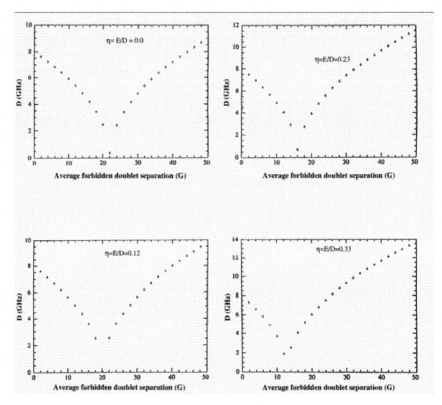

Figure 2. Plots of values of $|D|$ as functions of the average hyperfine forbidden doublet separation in the central sextet for the Mn(II) ion for: (a) η = 0.0, (b) η = 0.12, (c) η = 0.23, and η = 0.33. Adapted with permission from Misra (1995).

Mn(II) central sextet for (a) η = 0.0; (b) η = 0.12; (c) η = 0.23; (d) η = 0.33. Further, taking into account the first term, which has the largest value, it is concluded that the hyperfine doublet separation — $\Delta B^{(PC)}(1/2, m)$ — as given by eq. (23) is positive unless the value of $D > D_0$ for which $\Delta B^{(PC)}(1/2, m) = 0$. It is also noted that, because of G_0^2 [$\approx (h\nu)^2$, where h and ν are Planck's constant and the klystron frequency, respectively] in the denominator of the first term in (23), the hyperfine doublet separation decreases with increasing microwave frequency; for example, it will be more than 12 times smaller at the Q-band (~35 GHz) than that at the X-band (~10 GHz) for the central sextet. Thus, X-band data are preferable over Q-band data to estimate D from forbidden hyperfine doublet separation. To this end, it should be noted that the lower the microwave frequency the greater the resolution of hyperfine doublet separation.

To illustrate the use of third-order perturbation expressions, simulated spectra at the X- and K-bands for the central hyperfine sextet are illustrated in Figures 3 and 4, respectively, for various values of D. In addition to the above, it is possible to use the relative intensities of various hyperfine lines in the EPR spectra to estimate D as illustrated by Allen (1965) and depicted in Figure 5.

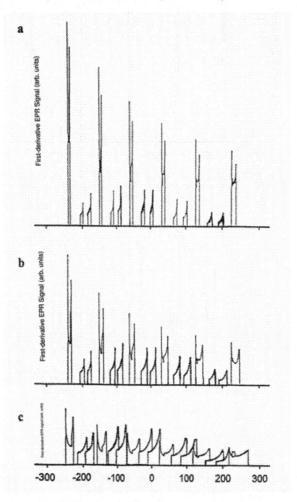

Figure 3. Computer-simulated spectra of the $M = +1/2 \leftrightarrow -1/2$ fine-structure group at the X band as calculated from expressions calculated up to third order in perturbation for various values of D. The abscissa is in gauss with the zero of reference chosen at unperturbed resonance field B_0. Full lines refer to $\Delta m = 0$ transitions, dashed lines to the hyperfine "forbidden" $\Delta m = \pm 1$ transitions. All intensities are drawn on the same scale. These spectra are to be compared with Figure 1b. $A/g\mu_B = 93$ G. (a) $D/g\mu_B = 75$ G, (b) $D/g\mu_B = 100$ G, and (c) $D/g\mu_B = 150$ G. Adapted with permission from de Wijn and Van Balderin (1967).

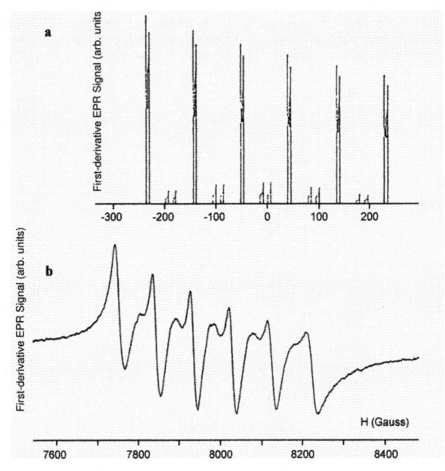

Figure 4. (a) Theoretical spectrum of the M = +1/2 ·· −1/2 fine-structure group at the K band. The abscissa is in Gauss with reference of zero at H_0. $A/g\mu_B$ = 93 G, $D/g\mu_B$ = 100 G. This is to be compared with Figure 4b, displaying the experimental first-derivative EPR spectrum of Mn(II) in a borate glass $(MnO)_x(B_2O_3+0.04K_2O)_{1-x}$ (x = 0.006) recorded at 23 GHz at 295 K. Adapted with permission from de Wijn and Van Balderin (1967).

3. SIMULATION OF POWDER SPECTRUM ON A COMPUTER USING MATRIX DIAGONALIZATION

In the brute-force technique, for each chosen orientation (θ,φ) of **B**, one diagonalizes the spin-Hamiltonian matrix for a large number of values of **B** to find the resonant magnetic-field value. It requires exorbitant computer time, especially

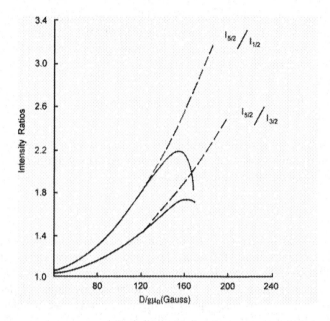

Figure 5. Ratios of line intensities versus $D/g\mu_B$ (Gauss). The solid lines show the second-order ratios calculated for D values up to $D/g\mu_B$ = 170 G; the dashed lines exhibit the extrapolated curve from the calculated curve up to $D/g\mu_B$ = 120 G. Adapted with permission from Allen (1965).

when the spin is large, *e.g.*, S = 5/2 for Mn(II). A technique of quickly calculating angular variation of resonant EPR line positions in a single crystal consists in calculating EPR line position by the method of least-squares fitting (LSF) for an orientation of **B** infinitesimally close to the one for which the resonant line position is known (Misra & Vasilopoulos, 1983). This can be easily extended to simulate a polycrystalline spectrum quickly. In the brute-force technique, for each chosen orientation of **B**, the spin-Hamiltonian matrix is diagonlized over a large number of closely spaced values of **B** to find the resonant magnetic field values. This requires exorbitant computer time, especially when the spin is as large as $S=5/2$ for the Mn(II) ion.

The EPR spectrum in a polycrystalline material can be simulated rigorously by overlapping spectra computed for a large number of orientations (θ,φ) of the external magnetic (Zeeman) field, **B**, over the unit sphere weighted in proportion to $\sin\theta \, d\theta \, d\varphi$ to take into consideration the distribution of various constituting crystallites whose principal axes are oriented in the interval $d\theta$, $d\varphi$ about (θ,φ). As well, a lineshape function, $F(\omega_i,B)$, for the various possible transitions $i' \leftrightarrow i''$, is used that could be Gaussian, Lorentzian, or a complicated function appropriate to the sample, in addition to each line position being weighted in proportion to its transition probability. The eigenvalues and eigenvectors required for calculation of line positions and line intensities are obtained by a numerical diagonalization of the spin-Hamiltonian matrix on a computer using the JACOBI subroutine (Press *et al.*, 1992). The following description is based on details given by Misra (1999).

The simulated spectrum can be expressed as

$$S(B,v) = \int_{\theta=0}^{\pi/2} \int_{\varphi=0}^{2\pi} \sum_{i} P(i,\theta,\varphi,v)F(\omega_i,B)d(\cos\theta)d\varphi \tag{25}$$

In (25) $P(i,\theta,\varphi,v)$ is the transition probability for the ith transition, between levels i' and i'', participating in resonance at microwave frequency v at orientation (θ,φ) of **B** over the unit sphere. It is expressed as follows:

$$P(i,\theta,\varphi) \propto \left|\left\langle \Phi_{i'}\left|(B_{1x}S_x + B_{1y}S_y + B_{1z}S_z)\right|\Phi_{i''}\right\rangle\right|^2 \tag{26}$$

In (26) S_α and $B_{1\alpha}$ ($\alpha = x,y,z$) represent the components of the electron spin operator, S, and modulation r.f. field B_1. $|\Phi_{i'}\rangle$ and $|\Phi_{i''}\rangle$ are the eigenvectors of the spin-Hamiltonian (SH) matrix, **H**, corresponding to energy levels E_i and E_i participating in resonance $[H|\Phi_k\rangle = E_k|\Phi_k\rangle$. For an arbitrary orientation of field B_1, one can sum the real and imaginary parts of the three terms together. The required square of the absolute value in (26) is as follows:

$$\left\langle \Phi_{i'}\left|B_{1x}S_x + B_{1y}S_y + B_{1z}S_z\right|\Phi_{i''}\right\rangle^2 = B_1^2 \left[\left|\sum_{\alpha=x,y,z} a_\alpha \, \text{Re}\left\{S_\alpha\left(|\Phi_{i''}\rangle \otimes \langle\Phi_{i'}|\right)\right\}\right|^2\right.$$
$$\left. + \left|\sum_{\alpha=x,y,z} a_\alpha \, \text{Im}\left\{S_\alpha\left(|\Phi_{i''}\rangle \otimes \langle\Phi_{i'}|\right)\right\}\right|^2\right] \tag{27}$$

where $a_x = \sin\theta\cos\varphi$, $a_y = \sin\theta\sin\varphi$, $a_z = \cos\theta$, Re and Im are the real and imaginary parts, respectively, and $\otimes$ represents the outer product of two column vectors. For use in (27), the following deductions are useful as far as the various terms are concerned.

Noting that

$$\text{Re}(S_x)_{j,j+1} = \text{Re}(S_x)_{j+1,j}, \text{ and } \text{Im}(S_x)_{j,k} = 0 \tag{28}$$

and that

$$\text{Re}\left(|\Phi_{i''}\rangle \otimes \langle\Phi_{i'}|\right)_{jk} = \text{Re}|\Phi_{i''}\rangle_j \, \text{Re}|\Phi_{i'}\rangle_k + \text{Im}|\Phi_{i''}\rangle_j \, \text{Im}|\Phi_{i'}\rangle_k$$
$$\text{Im}\left(|\Phi_{i''}\rangle \otimes \langle\Phi_{i'}|\right)_{jk} = -\text{Re}|\Phi_{i''}\rangle_j \, \text{Im}|\Phi_{i'}\rangle_k + \text{Im}|\Phi_{i''}\rangle_j \, \text{Re}|\Phi_{i'}\rangle_k \tag{29}$$

one obtains

$$\mathrm{Re}\,Tr\left[S_x\left(|\Phi_{i''}\rangle\otimes\langle\Phi_{i'}|\right)\right]=$$
$$\sum_j (S_x)_{j,j+1}\left\{\mathrm{Re}\left(|\Phi_{i''}\rangle\otimes\langle\Phi_{i'}|\right)_{j+1,j}+\mathrm{Re}\left(|\Phi_{i''}\rangle\otimes\langle\Phi_{i'}|\right)_{j,j+1}\right\} \tag{30}$$

$$\mathrm{Im}\,Tr\left[S_x\left(|\Phi_{i''}\rangle\otimes\langle\Phi_{i'}|\right)\right]=$$
$$\sum_j (S_x)_{j,j+1}\left\{\mathrm{Im}\left(|\Phi_{i''}\rangle\otimes\langle\Phi_{i'}|\right)_{j+1,j}+\mathrm{Im}\left(|\Phi_{i''}\rangle\otimes\langle\Phi_{i'}|\right)_{j,j+1}\right\} \tag{31}$$

Similarly, noting that

$$\mathrm{Im}(S_y)_{j,j+1}=-\mathrm{Im}(S_y)_{j+1,j}\quad\text{and}\quad \mathrm{Re}(S_y)_{jk}=0 \tag{32}$$

one obtains

$$\mathrm{Re}\,Tr\left[S_y\left(|\Phi_{i''}\rangle\otimes\langle\Phi_{i'}|\right)\right]=-\sum_j \mathrm{Im}(S_y)_{j,j+1}\,\mathrm{Im}\left(|\Phi_{i''}\rangle\otimes\langle\Phi_{i'}|\right)_{j+1,j} \tag{33}$$

$$\mathrm{Im}\,Tr\left[S_y\left(|\Phi_{i''}\rangle\otimes\langle\Phi_{i'}|\right)\right]=\sum_j \mathrm{Im}(S_y)_{j,j+1}\,\mathrm{Re}\left(|\Phi_{i''}\rangle\otimes\langle\Phi_{i'}|\right)_{j+1,j} \tag{34}$$

In (33) and (34) the required imaginary and real parts of the outer products on the right-hand sides are given by (29). As for the corresponding expression in S_z, one notes that S_z has only diagonal nonzero element, which is real, and thus,

$$\mathrm{Im}(S_z)_{jk}=0;\quad \mathrm{Re}(S_z)_{jk}=(S_z)_{jk}\,\delta_{jk} \tag{35}$$

where δ_{jk} is the Kronecker-delta symbol, such that $\delta_{ij}=0$ for $i\neq j$, $\delta_{ij}=1$ for $i=j$. Finally,

$$\mathrm{Re}\,Tr\left[S_z\left(|\Phi_{i''}\rangle\otimes\langle\Phi_{i'}|\right)\right]=$$
$$\sum_j (S_z)_{j,j}\left\{\mathrm{Re}|\Phi_{i''}\rangle_j\,\mathrm{Re}|\Phi_{i'}\rangle_j+\mathrm{Im}|\Phi_{i''}\rangle_j\,\mathrm{Im}|\Phi_{i'}\rangle_j\right\} \tag{36}$$

$$\mathrm{Im}\,Tr\left[S_z\left(|\Phi_{i''}\rangle\otimes\langle\Phi_{i'}|\right)\right]=$$
$$\sum_j (S_z)_{j,j}\left\{-\mathrm{Re}|\Phi_{i''}\rangle_j\,\mathrm{Im}|\Phi_{i'}\rangle_j+\mathrm{Im}|\Phi_{i''}\rangle_j\,\mathrm{Re}|\Phi_{i'}\rangle_j\right\} \tag{37}$$

The simulated spectrum is computed by the use of eq. (25), wherein the integrals are converted into discrete sums. It is clear from (25) that, in particular, one needs to know the resonant field values for the various transitions, as well as their transition probabilities for numerous orientations of the external magnetic field over the unit sphere over the unit sphere. A considerable saving of computer time can be accomplished if one uses numerical techniques to minimize the number of required diagonalizations of the SH matrix in the brute-force method. That is, when one uses the known resonant-field value at angle (θ,φ) to calculate the one at an infinitesimally close orientation, $(\theta+\delta\theta, \varphi+\delta\varphi)$, known as the method of homotopy (Misra & Vasilopoulos, 1980), as described below. The various parameters/ techniques required in the computation are described below.

Resonant Line Positions. The procedure to calculate the resonant line position at the orientation, $(\theta + \delta\theta, \varphi + \delta\varphi)$, from the knowledge of the line position at orientation (θ, φ), using the least-squares fitting technique and Taylor-series expansion, is as follows (Misra & Vasilopoulos, 1980; this paper describes the homotopy technique to calculate angular variation of EPR line positions.):

$$B_r(i, \theta+\delta\theta, \varphi+\delta\varphi) = \text{Iterative limit of}\left[-\left(\frac{\partial^2 S}{\partial B^2}\right)_{B_r}^{-1}\left(\frac{\partial S}{\partial B}\right)_{B_r}\right] \tag{38}$$

Thus, to be used in eq. (3.1),

$$-\left(\frac{\partial^2 S}{\partial B^2}\right)^{-1}\left(\frac{\partial S}{\partial B}\right) = -\left(\left|E_{i'} - E_{i''}\right| - h\nu\right) \times$$

$$\text{sign}\left(E_{i'} - E_{i''}\right)/\left(\frac{\partial E_{i'}}{\partial B} - \frac{\partial E_{i''}}{\partial B}\right) \tag{39}$$

In (38) one starts with $B_r' = B_r(i, \theta, \varphi)$, and S is defined as

$$S \equiv \left(\left|E_{i'} - E_{i''}\right| - h\nu\right)^2 \tag{40}$$

In eq. (39) the derivative of eigenvalue $E_{i'}$ or $E_{i''}$ can be evaluated as follows:

$$\frac{\partial E_{i'}}{\partial B} = \left\langle\Phi_{i'}\left|\frac{\partial H}{\partial B}\right|\Phi_{i'}\right\rangle = \mu_B\left\langle\Phi_{i'}\left|g_{zz}\cos\theta S_z + g_{xx}\sin\theta\cos\varphi S_x + g_{yy}\sin\theta\sin\varphi S_y\right|\Phi_{i'}\right\rangle \tag{41}$$

where, $g_{\alpha\alpha}$ ($\alpha = x, y, z$) are the components of the $\tilde{g}$-tensor . (It has been assumed that the $\tilde{g}$-tensor is diagonal in the coordinate axes chosen here; thus, $g_{xx} = g_{yy} = g_{zz} = g$) In writing (41), the fact that only the Zeeman term, $H_z = \mu_B \mathbf{S} \cdot \tilde{g} \cdot \mathbf{B}$, in the spin-Hamiltonian depends on the external field, and $B_z = B\cos\theta$, $B_x = B\sin\theta\cos\varphi$, $B_y = B\sin\theta\sin\varphi$, have been taken into account. There are, of course, other field-dependent terms, commonly not used, present in the spin-Hamiltonian which depend on higher powers of B, e.g., B^3, B^5,..., or on higher powers of S, e.g., S^3, S^5,... (see Misra et al., 1996; Al'tshuler & Kozyrev, 1974).

Transition Probabilities. The transition probability, given by eq. (26), at the infinitesimal orientation $\theta + \delta\theta$ and $\varphi + \delta\varphi$ of **B** can be obtained from eigenvectors $\left|\Phi_i(\theta+\delta\theta, \varphi+\delta\varphi)\right\rangle$. The latter can be calculated using $B_r(i, \theta+\delta\theta, \varphi+\delta\varphi)$ as obtained using eq. (38), and diagonalizing the spin-Hamiltonian expressed at $(\theta + \delta\theta, \varphi + \delta\varphi)$ for this value of the magnetic field.

Integrals. The integral for polycrystalline spectrum $S(B, \nu_c)$ as given by eq. (25) can be expressed as a sum over different orientations (θ_j, φ_j) of **B** distributed over the unit sphere divided into grids whose intersections for successive grids are infinitesimally close to each other, and over the values of **B** divided into channels,

B_k, distributed over the range of the magnetic field considered. Thus, eq. (25) can be expressed, using constant C, as the following sum:

$$S(B,\nu) = C \sum_{i,\theta_j,\varphi_j,k} P(i,\theta_j,\varphi_j,\nu) F(\omega_i, B_r(i,\theta_j,\varphi_j,\nu), B_k) \sin\theta_j \qquad (42)$$

where the values of θ_j are distributed over the range 0 to $\pi/2$, while those of φ_j over 0 to 2π, taking into account the fact that the EPR spectrum remains unchanged when the magnetic field orientation is reversed in direction due to time-reversal invariance; and $\sin\theta_j$ takes into account the uniform distribution of the crystallites constituting the powder such that the number of crystallites with their axes along θ_j is proportional to $\sin\theta_j$. Alternatively, the presence of factor $\sin\theta_j$ can be easily seen to arise, since solid angle $d\Omega = \sin\theta \, d\theta \, d\varphi$ represents the surface of the unit sphere covered by the orientation of **B** in the angular interval θ and $\theta + d\theta$, φ and $\varphi + d\varphi$. When the integration over the unit sphere is carried out as mentioned above to compute a powder spectrum, the various crystallites in the polycrystalline sample are fully taken into account. (The choice of grid (θ_j,φ_j) is described below.) Summation over k takes into account the probability of amplitude of absorption at the magnetic field value B_k due to lineshape distribution $F(\omega_i, B_r(i,\theta_j,\varphi_j,\nu), B_k)$ for the ith transition for the orientation of **B** along the (θ_j,φ_j) direction.

The (θ_j,φ_j) Grid. One can conveniently choose a (θ_j,φ_j) grid where the value of θ changes from 0 to $\pi/2$ in n_θ, steps where n_θ is a sufficiently large number, say, $n_\theta = 300$ (*i.e.*, every $3/10^{th}$ of a degree), depending upon the convergence of $B_r(i,\theta_j,\varphi_j,\nu)$ values computed by the use of eq. (27). Similar considerations apply to changes in φ values in n_φ steps, say, $n_\varphi = 120$ (*i.e.*, every 3 degrees, or even every degree). When there appear "crossing" or "looping" transitions, *e.g.*, in the case of the Fe^{3+} ion (Misra & Vasilopoulos, 1983), problems arise when two transitions cross each other between two successive (θ_j,φ_j) values considered ("crossing" transition), or a transition does not occur at the adjacent (θ_j,φ_j) values ("looping" transition). In order to overcome these, certain strategies may be employed as discussed by Misra and Vasilopoulos (1983). In addition, an improved partitioning scheme of the grid may be used. To this end, Wang and Hanson (1995, 1996) developed a novel scheme, named the SOPHE (Sydney Opera House) partitioning scheme, involving a combination of cubic spline and linear interpolations; the unit sphere is partitioned into triangularly shaped convexes subtending nearly the same solid angles; see also partitioning schemes proposed earlier *e.g.*, the Igloo method for partitioning an octant of the unit sphere (Nilges, 1979; Belford & Nilges, 1979; Maurice, 1980).

Lineshape Function $F(\omega_i, B_r(i,\theta_j,\varphi_j,\nu),B_k)$. The spectrum is then calculated by performing the summation in eq. (42) with $P(i,\theta_j,\varphi_j)$ centered at $B_r(i,\theta_j,\varphi_j,\nu)$ with lineshape function $F(\omega_i, B_r(i,\theta_j,\varphi_j,\nu))$ extended over a reasonable magnetic-field interval $\pm\Delta B$ about $B_r(i,\theta_j,\varphi_j,\nu)$, characteristic of the lineshape. The most commonly used lineshapes are Gaussian $[\sim\exp(B - B_0)^2/\sigma^2]$ and Lorentzian $[\sim\Gamma/\{(B - B_0)^2 + \Gamma^2\}]$, where σ and Γ are Gaussian and Lorentzian linewidths, respectively, and B_0 is the peak position. For computational purposes they can be expressed as follows:

Gaussian lineshape:

$$F_G(B_k, B_{ri}) = K_G \exp\left[-(B_k - B_r(i, \theta_j, \varphi_j, v_c))^2 / \sigma^2\right], \tag{43}$$

where B_r is the resonant field value for the ith transition, and K_G is a normalization constant.

Lorentzian lineshape:

$$F_L(B_k, B_r) = K_L \Gamma\left[\Gamma^2 + (B_k - B_r(i, \theta_j, \varphi_j, v_c))^2\right]^{-1}, \tag{44}$$

where $\Gamma = \sqrt{3}\Delta W/2$, where ΔW is the half-width at half-maximum, HWHM.

Computation of Eigenvalues and Eigenvectors of the Spin-Hamiltonian Matrix. They are computed by the use of JACOBI subroutine (Misra, 1999), which diagonalizes real symmetric matrices, and is particularly efficient when the off-diagonal elements in the SH matrix are infinitesimally small as is naturally the case in homotopy. (Briefly, the diagonalization in the JACOBI algorithm is accomplished by successive rotations to annihilate the off-diagonal elements of the 2×2 submatrix constituted by the largest off-diagonal element and corresponding diagonal elements of the SH matrix at any stage of successive rotations. For an efficient algorithm, see Press *et al.* (1992) and Misra (1999a).

Absorption Signal. Usually, it is the first derivative of the absorbed microwave power that is experimentally measured. The simulated first-derivative spectrum is calculated by taking the derivative with respect to **B** of $S(B, v)$, as given by eq. (42), along with that of the lineshape. Specifically, for the Lorentzian lineshape, given eq. (44), one has for the first-derivative

$$\partial F_L(\omega_i, B_r(i, \theta_j, \varphi_j, v), B_k)/\partial B_k =$$
$$-2K_L \Gamma[\Gamma^2 + (B_k - B_{ri})^2]^2 (B_k - B_{ri}) \tag{45}$$

Thus, the simulated first-derivative absorption spectrum is expressed, from (42), as

$$\partial S(B, v_c)/\partial B_k = C \sum_{i, \theta_j, \varphi_j, k} P(i, \theta_j, \varphi_j), \; \partial F_L(\omega_i, B_r(i, \theta_j, \varphi_j, v), B_k)/\partial B_k \sin \theta_j \tag{46}$$

$$= N \sum_{i,k} |\langle \Phi_{i'} | B_{1x} S_x + B_{1y} S_y + B_{1z} S_z | \Phi_{i'} \rangle|^2 (B_k - B_{ri})[\Gamma^2 + (B_k - B_{ri})^2]^{-2} \tag{47}$$

In (47) normalization constant N may be appropriately chosen, *e.g.*, the calculated value with the largest magnitude of the y value of all the channels was here set equal to 1.

The simulation procedure presented here, among others, is eminently exploitable to estimation of spin-Hamiltonian parameters from a powder spectrum by the least-squares fitting (LSF) procedure in conjunction with numerical diagonalization of the spin-Hamiltonian matrix similar to that proposed by Misra (1976) in context with single-crystal EPR spectra, even though the powder spectrum possesses further complications over and above those of a single-crystal EPR spectrum. This LSF technique will, thus, be of immense help to this end, especially in those cases where single-crystal samples cannot be prepared, *e.g.*, transition metal ion-doped metalloproteins. Efforts are currently in progress to accomplish this (Misra, 1999b).

4. COMPUTER SIMULATION OF Mn(II) EPR SPECTRA IN AMORPHOUS MATERIALS

Amorphous materials, *e.g.*, glasses or biological systems, are characterized by structure disorders, subjecting Mn(II) ions to experience random surroundings. The spin-Hamiltonian parameters thus vary rather widely. In powders the interpretation of EPR spectra becomes difficult due to the orientational disorder of paramagnetic ions. There are three types of glasses in which Mn(II) spectra have been mainly studied: borate (Misra, 1996, 1999), phosphate and silicate (Kliava & Purans, 1980; see Misra, 1996, 1999; de Wijn & Van Balderin, 1967, for more examples). Some materials pass from glassy to disordered state with variation of temperature as illustrated in Figure 6 (Allen, 1965).

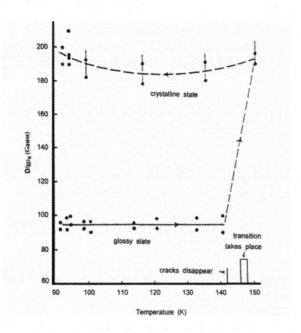

Figure 6. Variation of the value of $D/g_B{}^*$ of the Mn(II) ion in 12 N HCl with temperature. The arrows show the direction of the temperature changes in the experiment. Adapted with permission from Allen (1965).

Simulation of EPR spectra in glasses requires use of rather precise Mn(II) EPR lineshapes taking into account distribution of spin-Hamiltonian parameters [5]. Only the parameter values restricted to $|D| \ll g\mu_B B$, $|E| \ll g\mu_B B$, $|A| \ll g\mu_B B$ will here be considered. The resonance magnetic fields for transitions between states M, m and $M{-}1$, $m{+}i$, denoted by

$$B_0 = B_{M,m;M-1,m+i}(g,A,D,E,\theta,\varphi)$$

have been calculated (de Wijn & Van Balderin, 1967; Bleaney & Rubins, 1961; Bir, 1964, Upreti, 1974). In the expression for B_0, $i=0$ implies allowed, while $i=\pm 1, \pm 2, \ldots$ correspond to forbidden transitions; θ, φ are the polar and azimuthal angles formed by the magnetic field with the principal axes of the second-order crystal-field tensor in the spin-Hamiltonian. Finally, a total of nine parameters are to be determined: g_0, A_0, D_0, E_0, Δg, ΔA, ΔD, ΔE, and ΔB_{pp}, where subscript 0 denotes mean values, each of which is characterized by a width — Δg, ΔA, ΔD, ΔE — along with the peak-to-peak linewidth, ΔB_{PP}. Parameters g and A for Mn(II) are much less sensitive to change in the environment in comparison to the zero-field splitting fine-structure parameters, D and E. Thus, Δg and ΔA are relatively much smaller, and their variation can be taken into account by an appropriate value of ΔB_{PP}, which also takes into account spin–spin interaction and other terms not included in the spin-Hamiltonian. The value of $g_0 = 2.0$ can be safely assumed for the S-state ion Mn(II).

Considering only the variations ΔD and ΔE to be significant, describing Gaussian spreads with variances $\Delta D^2/2$ and $\Delta E^2/2$, respectively, and correlation coefficient r ($-1 \le r \le 1$) the joint statistical probability then becomes

$$P(D,E) = \left(\pi \Delta D \Delta E \sqrt{(1-r^2)}\right)^{-1} \times$$
$$\exp\left\{-\frac{1}{(1-r^2)}\left[\left(\frac{D-D_0}{\Delta D}\right)^2 - 2r\frac{(D-D_0)(E-E_0)}{\Delta D \Delta E} + \left(\frac{E-E_0}{\Delta E}\right)^2\right]\right\} \tag{48}$$

When $r = 0$, D and E are totally uncorrelated; $P(D,E)$ is expressed as the product of two Gaussian distribution that are independent of each other:

$$P(D) = \frac{1}{\sqrt{\pi}\Delta D}\exp\left[-\left(\frac{D-D_0}{\Delta D}\right)^2\right];$$

$$P(E) = \frac{1}{\sqrt{\pi}\Delta E}\exp\left[-\left(\frac{E-E_0}{\Delta E}\right)^2\right] \tag{49}$$

Values $r = \pm 1$ represent a total correlation between D and E. $P(D,E)$ as given by (48) is thus nonzero only when $(E - E_0)/(D - D_0) = \pm\Delta E/\Delta D$, in which case only one variation — $P(D)$ or $P(E)$ — is to be used in (49).

Assuming that ensembles of randomly oriented identical sites and randomly distorted sites are mutually independent, the EPR spectra in a glassy material can be expressed taking into account all fine-structure transitions and all orientations of Mn(II) ions, following Taylor and Bray (1972), as

$$P(H) = \sum_{\substack{m=-3/2 \\ i \\ |m-i| \le 5/2}} \sum_{m=-5/2}^{5/2} \int_{-\infty}^{\infty} dD \int_{-\infty}^{\infty} dE \int_{0}^{\pi} d\theta \int_{0}^{2\pi} d\varphi$$

$$\times P(D,E) \bullet \sin\theta \bullet W_{M,m,M-1,m+i}(D,E,\theta,\phi) \bullet F\left(\frac{B-B_0}{\Delta B_{PP}}\right)$$

(50)

In expression (50), the term $F[(B - B_0)/\Delta B_{PP}]$ is a lineshape function, and it can be considered the first derivative of either a Gaussian or a Lorentzian; $W_{M,m;M-1,m+i}(D,E,\theta,\phi)$ is the probability of a transition between states M, m and $M-1$, $m+i$ averaged over different orientations of the microwave magnetic field. W can be calculated using the method of Bir (1964), which is superior to the traditional perturbation approach (Bleaney & Rubins, 1961; as described in detail by Kliava & Purans, 1978) unless exact eigenvectors are used as obtained by computer diagonalization of the spin-Hamiltonian matrix as described in §3. The limits of integration over D and E can be taken to be $D_0 \pm 2\Delta D$ and $E_0 \pm 2\Delta E$. Finally, factor $\partial B_0/\partial h\nu$, which should be included in the integrand of (50), has here been approximated to be unity as its value deviates, in fact, from unity only by a few percent (Aasa & Vangard, 1975).

In practice, the calculation is confined to the transitions belonging to the central hyperfine sextet 1/2, $m \leftrightarrow -1/2$, $m + i$ (m = 5/2, 3/2, 1/2, –1/2, –3/2, —5/2), since the non-central lines are much less intense as compared to the central lines because they have a stronger angular dependence. In fact, in glasses parameter distributions totally smear out all non-central transition lines since their resonant fields contain terms linear in D and E, which is not the case for the resonant fields for the central transition lines.

5. COMPUTER-SIMULATED SPECTRA AND COMPARISON WITH EXPERIMENT

Figures 7–14 display various simulated and experimental spectra. The salient features of these are as follows:

(i) Figure 8 indicates a drastic discrepancy between the two simulated spectra, one which does and the other which does not take into account different transition probabilities of different transitions in the simulated spectrum, leading to the conclusion that transition probabilities are extremely important in arriving at a reasonable fit.

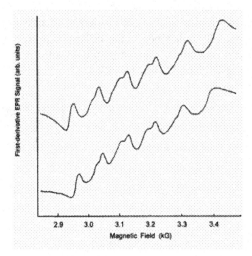

Figure 7. EPR spectra of Mn(II) observed at $\nu = 8.9$ GHz in phosphate (ZnO·P$_2$O$_5$, upper trace) and silicate (K$_2$O·4Si$_2$O, lower trace) glasses. Adapted with permission from Kliava and Purans (1980).

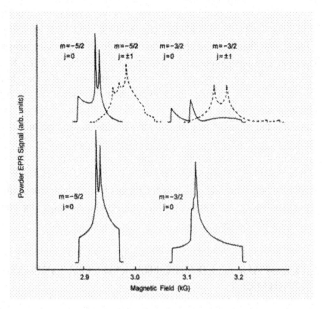

Figure 8. Mn(II) powder spectra simulated to third order in perturbation for some allowed (solid lines) and forbidden (dashed lines) hyperfine transitions belonging to the central fine-structure transition: $g = 2.0$, $A/g\mu_B = -95$ G, $D/g\mu_B = 210$ G, $E/g\mu_B = 70$ G, $\nu = 8.9$ GHz. For the top diagram the transition probability is calculated using Bir's method (1964), while in the bottom diagram the transition probability is equal to 1. The difference in the two diagrams shows the importance of taking into account the transition probabilities in spectral simulation. Adapted with permission from Kliava and Purans (1980).

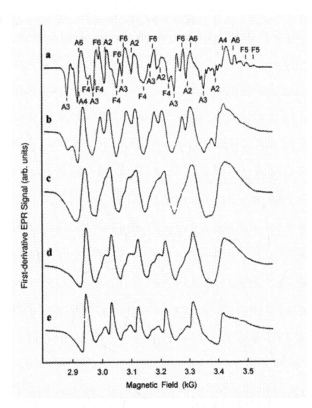

Figure 9. A series of spectra computed for different distributions of the fine-structure parameter (for the central fine-structure transition): $g = 2.0$, $A/g\mu_B = -93$ G, $D_0/g\mu_B = 220$ G, $E_0/g\mu_B = 73$ G, $\nu = 8.9$ GHz, $\Delta B_{PP} = 7$ G (Lorentzian lineshape function). The following values (in Gauss) of $\Delta D/g\mu_B$, $\Delta E/g\mu_B$ have been used for different traces: (a) 0, 0; (b) 40, 20; (c) 80, 40; (d) 120, 60; (e) 160, 80. In trace (a) the attribution of some of the spectral features to definite critical points is shown. Numbers 1–6 indicate the six types of critical points (see eq. (56)) for the allowed (A) and forbidden i = ±–1 (F) hyperfine-structure transitions. Adapted with permission from Kliava and Purans (1980).

(ii) Changing the signs of D_0 or E_0 does not lead to appreciable variation of simulated spectra since the third-order terms in resonant field linear in D_0 and E_0 are small.

(iii) The best-fit spin-Hamiltonian parameters are found to be the same for the two glasses from Figures 12 and 13 for phosphate and silicate glasses with the exception of A: $g = 2.0$, $|D_0|/g\mu_B = 220 \pm 20$ G, $|E_0|/g\mu_B = 70 \pm 15$ G, $\Delta D/g\mu_B = 80 \pm 20$ G, $\Delta E/g\mu_B = 30 \pm 10$ G, $r = 0.0 \pm 0.2$, while $A/g\mu_B = -93 \pm 1$ G for the phosphate glass and $A/g\mu_B = -87 \pm 1$ G for the silicate glass. Outside the limits of error for D_0, E_0, ΔD, ΔE, and A, the variation of one parameter degrades the other parameters. The limits of error are determined for correlation coefficient r for fixed values of all other parameters. Finally, it is noted that only the absolute values of

D_0 and E_0 can be determined in glasses when D_0 is small. On the other hand, when D_0 is large ($D_0/g\mu_B \geq 10,000$ G), then variable temperature measurements will allow determination of the signs for D, E. Zero-field EPR may also be used to determine the signs of D, E.

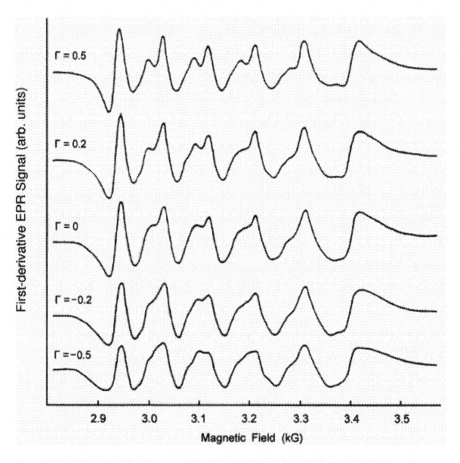

Figure 10. A series of spectra computed for different values of the central fine-structure transition: $g = 2.0$, $A/g\mu_B = -93$ G, $D_0/g\mu_B = 220$ G, $\Delta D/g\mu_B = 80$ G, $\Delta E/g\mu_B = 30$ G, $v = 8.9$ GHz, $\Delta B_{PP} = 6$ G (Lorentzian lineshape). Adapted with permission from Kliava and Purans (1980).

5.1. Effect of Distribution of Fine-Structure Parameters D and E on the Shape of EPR Spectra

The effect of the distributions of the fine-structure parameter on the shape of EPR spectra is illustrated in Figure 9. The "lines" and "peaks" that appear in the

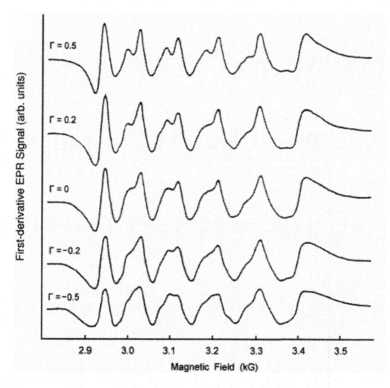

Figure 11. A series of spectra computed for different values of correlation coefficient r for the central fine-structure transition: $g = 2.0$, $A/g\mu_B = -93$ G, $D_0/g\mu_B = 220$ G, $E_0/g\mu_B = 73$ G, $\Delta D/g\mu_B = 80$ G, $\Delta E/g\mu_B = 30$ G, $\nu = 8.9$ GHz, $\Delta B_{PP} = 6$ G (Lorentzian lineshape). Adapted with permission from Kliava and Purans (1980).

absence of distribution of D and E (curve a) broaden at first as ΔD and ΔE are increased (curves b–e); some of them broaden out completely. As ΔD and ΔE are increased further, the remaining "lines" and "peaks" do not broaden any more, contrary to expectation, and even become narrower (curves d–e). This can be explained by expressing (50) as follows (Kneubuhl, 1960):

$$P(B) = \sum_{M,i,m} \int dB_0 \int_S d\sigma \cdot p(d,e) \cdot W_{M,m;M-1,m+i}(d,e,\theta,\varphi) \cdot$$
$$F\left(\frac{B - B_0}{\Delta B_{PP}}\right) / |\text{grad } B_0| \tag{51}$$

In expression (51) the integration is over B_0, three-dimensional surface S is defined by $B_0(d,e,\theta,\varphi) = $ constant, $d = D/\Delta D$ and $e = E/\Delta E$,

$$P(d,e) = \frac{1}{\pi \left(1 - r^2\right)^{1/2}} \times$$

$$\exp\left\{-\frac{1}{(1-r^2)}\left[(d-d_0)^2 - 2r(d-d_0)(e-e_0) + (e-e_0)^2\right]\right\} \tag{52}$$

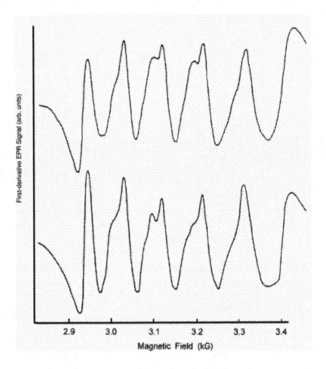

Figure 12. Comparison of the experimental and computer-simulated spectra for zinc-phosphate glass ($ZnO \cdot P_2O_5$). The upper trace represents the experimental spectrum after subtracting the broad underlying resonance. The lower trace is the computed best-fit spectrum (for the central fine-structure transition) with $g = 2.0$, $A/g\mu_B = -93$ G, $D_0/g\mu_B = 220$ G, $E_0/g\mu_B = 70$ G, $\Delta D/g\mu_B = 80$ G, $\Delta E/g\mu_B = 30$ G, $\nu = 8.9$ GHz and $\Delta B_{PP} = 6$ G (Lorentzian lineshape). Adapted with permission from Kliava and Purans (1980).

In eq. (52), $d_0 = D_0/\Delta D$, $e_0 = E_0/\Delta E$. Since H_0 depends on d, e, θ, and φ, one has

$$\left|\text{grad } B_0\right| = \left[\left(\text{grad}_{\theta,\varphi} B_0\right)^2 + \left(\frac{\partial B_0}{\partial d}\right)^2 + \left(\frac{\partial B_0}{\partial e}\right)^2\right]^{1/2}, \tag{53}$$

where

$$\left| \text{grad}_{\theta,\varphi} B_0 \right| = \left[\left(\frac{\partial B_0}{\partial \theta} \right)^2 + \frac{1}{\sin^2 \theta} \left(\frac{\partial B_0}{\partial \varphi} \right)^2 \right]^{1/2}. \tag{54}$$

When $\Delta D = \Delta E = 0$, it is seen from (51) that the "lines" and "peaks" in the powder EPR spectra occur at resonance-field values for which $\left| \text{grad}_{\theta,\varphi} B_0 \right| = 0$, *i.e.*,

$$\frac{\partial B_0}{\partial \theta} = 0, \quad \frac{1}{\sin \theta} \frac{\partial B_0}{\partial \varphi} = 0. \tag{55}$$

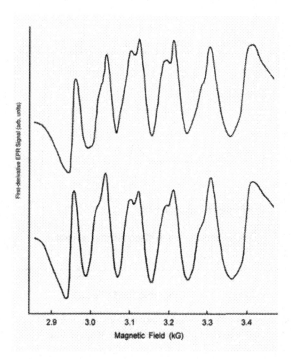

Figure 13. Comparison of the experimental and computer-simulated spectra for potassium silicate glass ($K_2O \cdot 4SiO_2$). The upper trace represents the experimental spectrum after subtracting the broad underlying resonance. The lower trace is the computed best-fit spectrum (for the central fine-structure transition) with $g = 2.0$, $A/g\mu_B = -87$ G, $D_0/g\mu_B = 220$ G, $E_0/g\mu_B = 70$ G, $\Delta D/g\mu_B = 80$ G, $\Delta E/g\mu_B = 30$ G, $\nu = 8.9$ GHz, $\Delta B_{PP} = 7$ G (Lorentzian lineshape). Adapted with permission from Kliava and Purans (1980).

It can be shown that for each hyperfine transition six different types of critical points occur at definite values of θ_0, φ_0, which are as follows:
1. $\theta_0 = 0$;
2, 3. $\theta_0 = \pi/2$, $\varphi = \pi/4 \mp \pi/4$ (or $5\pi/4 \mp \pi/4$);
4. $\theta_0 = \pi/2$, $\cos 2\varphi_0 = \alpha D/E$, where

$$\alpha = \frac{2 - (2m+i)A/B_0 + 3(2m+i)(1+i)A^2/DB_0}{18 - 73(2m+i)A/B_0}$$

5, 6. $\cos^2 \theta_0 = x$, $\varphi_0 = \pi/4 \mp \pi/4$ (or $5\pi/4 \mp \pi/4$), where

$$x = \frac{10D \pm 6E - (2m+i)(37D \pm 35E)A/B_0 - 3(2m+i)(1+i)A^2/B_0}{[18 - 73(2m+i)A/B_0](D \pm E)} \quad (56)$$

Assignment of some "lines" and "peaks" to define critical points is illustrated in Figure 9 (curve a).

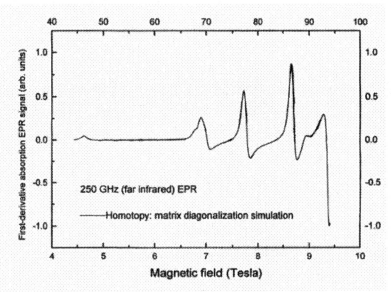

Figure 14. Simulated Mn(II) spectrum at 249.9 GHz with parameter values $g = 2.00$, $D_0/g\mu_B$ = 10,000 G, $E_0/D_0 = 0$. Reprinted with permission from Misra (1999a).

Supposing now that $\Delta D \neq 0$, $\Delta E \neq 0$. If the parameter distributions are small enough to satisfy the conditions

$$\left|\frac{\partial B_0}{\partial D}\right|_c \cdot \Delta D, \ \left|\frac{\partial B_0}{\partial E}\right|_c \cdot \Delta E << \frac{\partial}{\partial \theta}|grad_{\theta,\varphi}B_0|_c, \ \frac{\partial}{\partial \varphi}|grad_{\theta,\varphi}B_0|_c \quad (57)$$

where index c means that all the distributions are taken at the critical point (θ_0, φ_0, D_0, E_0), the value of $\partial B_0/\partial \theta$ and $\partial B_0/(\sin \theta \cdot \partial \varphi)$ increase rapidly as one moves away from the critical point. Finally, eq. (53) becomes

$$\left| \text{grad } B_0 \right|_c = \left[\left(\frac{\partial B_0}{\partial D} \right)_c^2 (\Delta D)^2 + \left(\frac{\partial B_0}{\partial E} \right)_c^2 (\Delta E)^2 \right]^{1/2} \tag{58}$$

Equation (58) reveals that the amplitudes of the "lines" and "peaks" decrease, while their widths increase, in proportion to $\left| \text{grad } B_0 \right|_c$. This broadening becomes significant when it approaches or exceeds the value of ΔB_{PP}, which includes all other sources of line broadening.

5.2. Sharp Features in Mn(II) Spectra

The salient features of the computer-simulation of sharp spectral features in the spectra in the region $g \approx 2.0$ are as follows:

(i) The values $D, E \approx 0$ account for sharp features in computer-simulated spectra at broad parameter distributions (Figure 9, curves d and e).

(ii) For the case of intermediate parameter distributions, no definite statement can be made on parameter distribution without a complicated analysis.

(iii) As for the dependence of the spectra on the ratio E_0/D_0, even at rather broad parameter distributions ($\Delta D/g\mu_B = 80$ G, $\Delta E/g\mu_B = 30$ G), an increase in E_0/D_0 produces a strong variation in spectral shape. Figures 14 and 15 represent 250-GHz spectra for two ratios of $E_0/D_0 = 0$ and $E_0/D_0 = 0.23$. It is seen from these figures that when E_0/D_0 is large, there is an apparent splitting of the $-2D$ and $-4D$ fine-structure lines (Wood et al., 1999). Computer fitting to the experimental spectra enables determination of the absolute value of ratio E_0/D_0 along with other spin-Hamiltonian parameters uniquely for Mn(II) in glasses with sufficient accuracy. Great care must be taken to "qualitatively" interpret EPR spectra of Mn(II) in disordered systems. For example, a spectrum with broad parameter distributions can be easily mistaken for one with small values of D_0 and E_0 (Taylor & Bray, 1972), since there are present some sharp features in both cases. This mistake is not possible with computer simulation of EPR spectra, since the overall features of computed spectra are quite different in the two cases.

5.3. Broad Resonances in Mn(II) Spectra

Computer simulations support the claim of Griscom and Griscom (1960) that the broad resonance does not arise from non-central fine-structure transitions, since the contribution of these transitions to the total EPR spectrum is, in fact, insignificant to account for the observed background resonance (see Figure 16.) As seen from Figure 7, the peak-to-peak intensities of the broad resonance relative to sharp central features are about 3.3 and 4.0, respectively, for the phosphate and silicate glasses.

Griscom and Griscom (1960) concluded from the fact that broad underlying resonances observed at X-band collapse at the K band, and that a broad distribution of sites with values $|D|/h$ ranging from near zero to as high as 7 GHz (~2.5 kG) and $|E_0/D_0| \approx 1/3$ is responsible for this behavior; sites with $|D|/h \geq 2$ GHz (~0.7

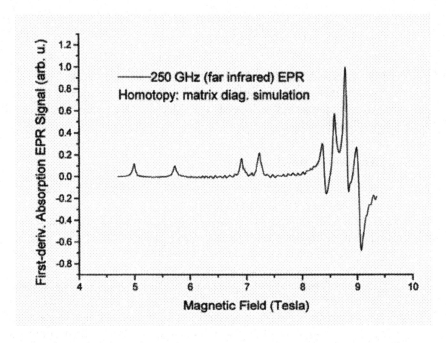

Figure 15. Simulated Mn(II) spectrum at 249.9 GHz with parameter values g = 2.00, $D_0/g\mu_B$ = 10,000 G, E_0/D_0 = 0.23. Reprinted with permission from Misra (1999a).

kG) give rise to background resonances. Computer simulations (Kliava & Purans, 1980), however, yield sharp g = 2.0 features with $|D_0|/g\mu_B \approx 220G$ (~0.6 GHz) with the same ratio $|E_0/D_0| \approx 1/3$, but without continuous broad distribution of sites. Thus, it is concluded that in addition to the large site-to-site distortion range, sites for which short-range order is much better preserved do exist in glasses (Tucker, 1962).

6. ESTIMATION OF SPIN-HAMILTONIAN PARAMETERS AND LINEWIDTH FROM A POWDER SPECTRUM AND CALCULATION OF FIRST AND SECOND DERIVATIVES OF THE χ^2-FUNCTION

This section deals with, in outline form, the essentials of the the least-squares fitting (LSF) technique; more details are given by Misra (1999a). For application of the LSF technique to evaluate spin-Hamiltonian parameters and linewidth, the χ^2-function is defined as the sum of the weighted squares of the differences of the calculated and measured first-derivative absorption signals at the magnetic field values B_k within the range of the magnetic field considered:

$$\chi^2 = \sum_k [F_c(B_k, v_c) - F_m(B_k, v_c)]^2 / \sigma_k^2,$$ (59)

where $F_c(B_k, v_c)$ and $F_m(B_k, v_c)$ are, respectively, the normalized calculated [eqs. (46) or (47)] and measured values of the first-derivative EPR signal, and σ_k is the weight factor (related to standard deviation of datum k). The measured/calculated values may be normalized in such a way that the maximum of each is equal to 1.

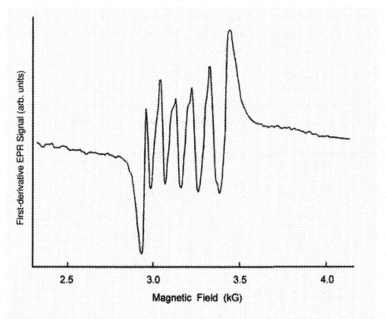

Figure 16. A spectrum computed taking into account all the fine-structure transitions: $g = 2.0$, $A/g\mu_B = -93$ G, $D_0/g\mu_B = 220$ G, $E_0/g\mu_B = 70$ G, $\Delta D/g\mu_B = 80$ G, $\Delta E/g\mu_B = 30$ G, $v = 8.9$ GHz and $\Delta B_{PP} = 11$ G. Adapted with permission from Kliava and Purans (1980).

In the LSF technique vector $\mathbf{a}^m$ constituted by the parameters that correspond to the absolute minimum of the χ^2-function can be obtained from $\mathbf{a}^i$, the vector constituted by the initially chosen value of SH parameters and linewidth by the following relation [the article by Misra (1976) deals with the LSF technique as applied to evaluate spin-Hamiltonian parameters from single-crystal EPR line positions]:

$$\mathbf{a}^m = \mathbf{a}^i - (\mathbf{D}''(\mathbf{a}^i))^{-1} \mathbf{D}'$$ (60)

In eq. (60) $\mathbf{D}'$ is the column vector whose elements are the first derivatives of the χ^2-function with respect to the parameters evaluated at $\mathbf{a}^i$ and $\mathbf{D}''$ is the matrix whose elements are the second derivatives with respect to the parameters evaluated at $\mathbf{a}^m$:

$$\mathbf{D}_m^{\odot} = \left(\frac{\partial \chi^2}{\partial a_m} \right)_{\mathbf{a}^i} \tag{61}$$

$$\mathbf{D}_{nm}^{''} = \left(\frac{\partial^2 \chi^2}{\partial a_n \partial a_m} \right)_{\mathbf{a}^m} \tag{62}$$

Since $\mathbf{a}^m$ is not known to begin with, the elements of matrix $\mathbf{D}''$ are, in practice, evaluated with respect to $\mathbf{a}^i$, referred to as $\mathbf{D}''$ $(\mathbf{a}^i)$. A new set of parameters, denoted by vector $\mathbf{a}^f$ is then calculated as follows:

$$\mathbf{a}^f = \mathbf{a}^i - (\mathbf{D}''(\mathbf{a}^i))^{-1} \mathbf{D}' \tag{63}$$

in place of vector $\mathbf{a}^m$ given by (60), which is calculated iteratively until a sufficiently small value of the χ^2-function, consistent with experimental uncertainties, is obtained.

Calculation of $\mathbf{D}'$ and $\mathbf{D}''$. From eqs. (61) and (62), one has

$$D_m^{'} = 2 \sum_k [F_c(B_k, \nu_c) - F_m(B_k, \nu_c)] \left(\frac{\partial F_c(B_k, \nu_c)}{\partial a_m} \right) \Big/ \sigma_k^2 \tag{64}$$

$$D_{nm}^{''} = 2 \sum_k \left\{ [F_c(B_k, n_c) - F_m(B_k, n_c)] \left(\frac{\partial^2 F_c(B_k, n_c)}{\partial a_n \partial a_m} \right) \right.$$
$$\left. + \left(\frac{\partial F_c(B_k, n_c)}{\partial a_n} \right) \left(\frac{\partial F_c(B_k, n_c)}{\partial a_m} \right) \right\} \Big/ \sigma_k^2 \tag{65}$$

Thus, in order to evaluate (64) and (65), one needs to calculate the first and second derivatives of $F_c(B_k, \nu_c)$, given by eq. (47), with respect to the parameters.

Finally, the parameter errors, Δa_j, are estimated statistically using the matrix constituted by the second derivatives of the χ^2-function (Misra & Subramanian, 1982):

$$\Delta a_j = \varepsilon_{jj}^{1/2} \tag{66}$$

where, matrix ε is defined as follows:

$$\varepsilon = (\tfrac{1}{2} \mathbf{D}'')^{-1}.$$

7. CONCLUDING REMARKS

The present work provides essential insights into Mn(II) EPR spectra in amorphous materials, appropriate to biological systems, and how to simulate and interpret them. It does not attempt to provide full details of the various publications on the subject. The many examples of Mn(II) EPR spectra in glasses provided here should serve as prototypes for biological systems, since glasses and biological systems exhibit similar spectra. A detailed list of references on EPR in metalloproteins has here been mentioned in §1. The various references (and references therein) should be helpful to obtain a rather exhaustive list of papers published on the subject. An interesting review article covering both NMR and EPR spectra in polycrystalline materials taking into account many different values of electronic and nuclear spins is provided by Taylor *et al.* (1975).

Mn(II) EPR spectra provide information about the environment of the Mn(II) ion in amorphous materials, which is reflected in the values of parameters g, D, and E and the linewidth. As for proteins with Mn(II) complexes, signal broadening due to motional effect or by structural variations diminishes as the composition of the complex approaches that of the fully functional state of the protein. Hence, the improvement in the resolution of the spectrum provides some insight into the fidelity of the structure under investigation. The interpretation of the structure is facilitated by studying the various glassy spectra provided in this article, since a close parallel of biological systems exists with analysis of other amorphous materials in the solid state.

8. REFERENCES

Aasa R, Vanngard T. 1975. *J Magn Reson* **19**:308.

Allen BT. 1965. *J Chem Phys* **43**:3820.

Al'tshuler SA, Kozyrev BM. 1974. Electron paramagnetic resonance in compounds of transition ions. New York: John Wiley.

Ash DE, Schramm VL. 1982. *J Biol Chem* **257**:9261.

Beinert H. 1985. *Biochem Soc Trans* **13**:542.

Belford RL, Nilges MJ. 1979. Computer simulation of powder spectra. Paper presented at EPR Symposium, 21st Rocky Mountain Conference, Denver, Colorado.

Berliner LJ, Ellis PD, Murakami K. 1983. *Biochemistry* **22**:5061.

Bir GL. 1964. *Sov Phys Solid State* **5**:1628.

Blanchard SC, Chasteen ND. 1976. *J Phys Chem* **80**:1362.

Bleaney B, Rubins RS. 1961. *Proc Phys Soc* (London) **77**:33; corrigendum, 1961, **78**:78.

Blumberg WE, Peisach J. 1987. *Life Chem Rep* **5**:5.

Boas JF. 1984. In *Copper proteins and copper enzymes*, Ed R Lontie. Boca Raton, FL: CRC Press, Boca Raton.

Chasteen ND. 1977. *J Phys Chem* **81**:1420.

Chien JCW, Dickinson LC. 1981. In *Biological magnetic resonance*. Ed LJ Berliner, J Reuben. New York: Plenum.

Chiswell B, McKenzie ED Lindboy LF. 1987. In *Comprehensive coordination chemistry*, Ed G Wilkinson, RD Gillard, JA McCleverty. Oxford: Pergamon, Oxford.

Coffino A, Pesiach J. 1996. *J Magn Reson* **B111**:127.

Coleman WM, Taylor LT. 1980. *Coord Chem Rev* **32**:1.

Crabbe MJC, Waight RD, Bardsley WG, Barker RW, Kelly ID, Knowles PF. 1976. *Biochem J* **155**:679.

Galbraith W, Goldstein IJ. 1970. *FEBS Lett* **9**:197.

Griscom DL, Griscom RE. 1967. *J Chem Phys* **47**:2711.

Hanson GR, Pilbrow JR. 1987. *Specialist periodical reports — electron spin resonance*, Vol 10B. Ed MCR Symons. London: The Royal Society of Chemistry.

Hanson GR, Wilson GL. 1988. *Specialist periodical reports — electron spin resonance*, Vol 11B, Ed MCR Symons. London: The Royal Society of Chemistry.

Hirsch-Kolb H, Kolb HJ, Greenber DM. 1971, *J Biol Chem* **246**:395.

Kliava JG, Purans J. 1978. *Phys Stat Solidi (a)* **49**:K43.

Kliava JG, Purans J. 1980. *J Magn Reson* **40**:33.

Kneubuhl FK. 1960. *Chem Phys* **33**:1074.

Leach Jr RM. 1971. *Fed Proc* **30**:991.

Markham GD, Rao BDN, Reed GH. 1979. *J Magn Reson* **33**:595.

Maurice AM. 1980. PhD dissertation. University of Illinois, Urbana.

McEuen AR. 1982. *Inorg Biochem* **3**:314.

Meirovitch E, Poupko R. 1978. *J Phys Chem* **82**:1920.

Misra SK. 1976. *J Magn Reson* **23**:403.

Misra SK. 1994. *Physica B* **203**:193.

Misra SK. 1996. *Appl Magn Reson* **10**:193.

Misra SK. 1997. *Physica B* **240**:183.

Misra SK. 1998. *Mol Phys Repts (Institute of Molecular Physics, Polish Academy of Sciences)* **26**:85.

Misra SK. 1999a. *J Magn Reson* **137**:83.

Misra SK. 1999b. *J Magn Reson* **140**:179.

Misra SK, Subramanian S. 1982. *J Phys C* **15**:7199.

Misra SK, Sun J. 1991. *Mag Reson Rev* **16**: 57.

Misra SK, Vasilopoulos PV. 1980. *J Phys C: Condens Matter* **13**:1083

Misra SK, Poole Jr CP, Farach HA. 1996. *Appl Magn Reson* **11**:29.

Musci G, Reed GH, Berliner LJ. 1986. *Inorg Biochem* **26**:229.

Nilges MJ. 1979. PhD dissertation. University of Illinois, Urbana, Illinois.

Palmer G. 1985. *Biochem Soc Trans* **13**:548.

Press WH, Teukolsky SA, Vetterling WT, Flannery BP. 1992. *Numerical recipes in fortran*, 2nd ed. New York: Cambridge UP.

Que Jr L, Widom J, Crawford RL. 1981. *J Biol Chem* **256**:10941.

Reed GH, Markham GD. 1984. In *Biological magnetic resonance*, Vol. 6. Ed LJ Berliner, J Reuben. New York: Plenum.

Smith TD, Pilbrow JR. 1980. In *Biological magnetic resonance*, Vol 2. Ed LJ Berliner, J Reuben. New York: Plenum.

Taylor PC, Bray PJ. 1972. *J Phys Chem Solids* **33**:43.

Taylor PC, Bangher JF, Kriz HM. 1975. *Chem. Rev.* **75**:203.

Tucker RF. 1962. *Advances in glass technology*. New York: Plenum.

Underwood EJ. 1971. *Trace elements in human and animal nutrition*, 3rd ed. New York: Academic, New York.

Upreti GC. 1974. *J Magn Reson* **13**:336.

Villafranca JJ, Raushel FM. 1990. *Adv Bio-Inorg Chem* **4**:289.

Wang D, Hanson GR. 1996. *Appl Magn Reson* **11**:401.

Wang D, Hanson GR. 1995. *J Magn Reson, Ser A* **117**:1.

Wedler FC, Denman RB, Roby WG. 1982. *Biochemistry* **21**:6389.

White LK, Szabo A, Crackner P, Sauer K. 1980. *Acct Chem Res* **13**:249.

de Wijn HW, Van Balderin RF. 1967. *J Chem Phys* **46**:1381.

Wood RM, Stucker DM, Jones LM, Lynch WB, Misra SK, Freed JH. 1999. *Inorg Chem* **38**:5384.

CHAPTER 5

DENSITY MATRIX FORMALISM OF ANGULAR MOMENTUM IN MULTI-QUANTUM MAGNETIC RESONANCE

H. Watari and Y. Shimoyama

Uchihonmachi 2-chome, Suita, Osaka, 564-0032 Japan; Department of Physics, Hokkaido University of Education, Hakodata, 040-0083 Japan

1. INTRODUCTION

The pulse Fourier transform approach to magnetic resonance spectroscopy has been extensively developed and successfully applied to systems of one-half spin and their mutual interactions. But resonance spectroscopy of spin systems with the higher half- and integer spin quantum numbers is commonplace, for example, in the case of alkali metal nuclear magnetic resonance (NMR) and electron paramagnetic resonance (EPR) of transition metal compounds involving multi-quantum transitions. Similarly, magnetic resonance at zero field entails the observation of multi-quantum transitions.

Besides the treatment of high-spin nuclei, multi-quantum transitions are becoming more frequently encountered because magnetic resonance spectroscopy has expanded beyond the traditional operating frequencies of microwave spectrometers. Magnetic resonance phenomena are being studied in the submillimeter region of the electromagnetic spectrum. And at the high-frequency end of the electromagnetic spectrum the Fourier transform is being applied to spectroscopy in ultraviolet, visual, and infrared regions in time-domain measurements using a Michelson interferometer and a birefingent interferometer. Additionally, a laser operating with a several-femtosecond pulsewidth has been developed and used to observe multi-quantum coherence effects in materials.

Magnetic and optical resonances are identical electromagnetic phenomena in the sense that there occurs an interaction of a magnetic field with matter, and both types of experiments may be described under a common mathematical formalism that is independent of experimental approach. But in developing a theoretical formalism that fuses both optical and magnetic resonance phenomena there occurs a problem reconciling the manner in which one treats the resonance condition. For example, both magnetic resonance theory and experiment deal directly with an "on-resonance" condition. But in quantum optics one describes a resonance condition via perturbation theory, and thus an application of the theory to the resonance condition has been limited to treatments in which the system is brought into near-resonance so as to avoid a divergence at the frequency of on-resonance. This fundamental difference between the two theoretical approaches is just one problem that must be addressed in reconciling the methodologies.

The density matrix representation of spin and orbital angular momentum is capable of expressing a static state of matter and its time-dependent response to an external perturbation. Our application necessitates that we follow the response of the orbital and spin momenta subject to full or partial excitations, and the density matrix provides a direct solution to the stochastic Liouville equation. But the density matrix representation in a rotating operator is algebraically ambiguious, and we must also clarify the algebraic description of selective excitation of multi-quantum systems.

The purpose of this paper is to present the density matrix formalism of angular momentum with half- and integer spin quantum numbers using the spectrum dissolving theorem. The density matrix formalism was developed for the laboratory and rotating frame in order to obtain a complete analytical representation for the spin excitation and response scheme. The density matrix contained in the rotation operator will become clear, as oppossed to the approximate treatment, and thereby from the theoretical process of off-resonance to that of just-resonance through the state of near-resonance will be visualized continuously.

2. PROJECTION OF ANGULAR MOMENTUM ON BASIS SETS

A matrix representation of Hamiltonian $\mathbf{H}$ is given by eq. (1) assuming eigenfunction matrix φ and eigenvalue matrix Λ:

$$\mathbf{H} \cdot \phi_{\mathbf{H}} = \phi_{\mathbf{H}} \cdot \Lambda_{\mathbf{H}} \tag{1}$$

where "·" denotes a matrix product. Given that $\mathbf{H}$, $\phi_{\mathbf{H}}$, and $\Lambda_{\mathbf{H}}$ are 3×3 square matrices, for convenience we can develop an algebraic representation for the larger quantum numbers. An eigenfunction matrix may be written as follows:

$$\phi_{\mathbf{H}} = \begin{vmatrix} p_p & q_p & r_p \\ p_q & q_q & r_q \\ p_r & q_r & r_r \end{vmatrix} \tag{2}$$

where a given element belongs to a proper eigenvalue, respectively, and the associated subscript refers to one of the connected basis set. An inverse of eigenfunction matrix $\phi_{\mathbf{H}}^{-1}$ is expressed as follows:

$$\phi_H^{-1} = \begin{vmatrix} p_p^* & p_q^* & p_r^* \\ q_p^* & q_q^* & q_r^* \\ r_p^* & r_q^* & r_r^* \end{vmatrix} \tag{3}$$

where an asterisk denotes a complex conjugate. Each element value in the eigen-value matrix can be obtained from the Hamiltonian **H** using the secular equation, and its expansion is also shown by matrices E_p, E_q, and E_r, which designate an individual basis set on the diagonal axis as

$$\Lambda_H = \begin{vmatrix} \lambda_p & 0 & 0 \\ 0 & \lambda_q & 0 \\ 0 & 0 & \lambda_r \end{vmatrix}$$

$$= \lambda_p \begin{vmatrix} 1 & 0 & 0 \\ 0 & 0 & 0 \\ 0 & 0 & 0 \end{vmatrix} + \lambda_q \begin{vmatrix} 0 & 0 & 0 \\ 0 & 1 & 0 \\ 0 & 0 & 0 \end{vmatrix} + \lambda_r \begin{vmatrix} 0 & 0 & 0 \\ 0 & 0 & 0 \\ 0 & 0 & 1 \end{vmatrix} \tag{4}$$

$$= \lambda_p E_p + \lambda_q E_q + \lambda_r E_r$$

$$= \sum_{j=p,q,r} \lambda_j E_j$$

Here λ_p, λ_q, and λ_r are nondegenerate eigenvalues. Among matrices E_p, E_q, and E_r, the following relation exists:

$$E = \sum_{j=p,q,r} E_j \tag{5}$$

$$E_j \cdot E_k = \delta_{jk} E_j \quad (j,k = p,q, \text{ and } r)$$

where δ is the Kronecker delta. Using ϕ_H, ϕ_H^{-1}, and Λ_H ..., the representation matrix can be resolved into a sum of individual terms as follows:

$$\mathbf{H} = \phi_H \cdot \Lambda_H \cdot \phi_H^{-1} = \sum_{j=p,q,r} \lambda_j \phi_H \cdot E_j \cdot \phi_H^{-1} \tag{6}$$

Each of the terms appearing in eq. (6) is defined as a projection matrix onto a basis set. Thus, each term $\phi_H \cdot E_j \cdot \phi_H^{-1}$ $(j = p,q,r)$ is defined as follows:

$$D_p = \phi_H \cdot E_p \cdot \phi_H^{-1} = \begin{bmatrix} p_p p_p^* & p_p p_q^* & p_p p_r^* \\ p_q p_p^* & p_q p_q^* & p_q p_r^* \\ p_r p_p^* & p_r p_q^* & p_r p_r^* \end{bmatrix}$$

$$D_q = \phi_{\mathsf{H}} \cdot E_q \cdot \phi_{\mathsf{H}}^{-1} = \begin{bmatrix} q_p q_p^* & q_p q_q^* & q_p q_r^* \\ q_q q_p^* & q_q q_q^* & q_q q_r^* \\ q_r q_p^* & q_r q_q^* & q_r q_r^* \end{bmatrix} \tag{7}$$

$$D_r = \phi_{\mathsf{H}} \cdot E_r \cdot \phi_{\mathsf{H}}^{-1} = \begin{bmatrix} r_p r_p^* & r_p r_q^* & r_p r_r^* \\ r_q r_p^* & r_q r_q^* & r_q r_r^* \\ r_r r_p^* & r_r r_q^* & r_r r_r^* \end{bmatrix}$$

When an eigenfunction is intrinsic, the projection matrix is itself a density matrix. Equation (7) then leads to

$$E = \sum_{j=p,q,r} D_j \tag{8}$$

where

$$D_j \cdot D_k = \delta_{ij} D_j \quad (j,k = p,q,r) \tag{9}$$

It is apparent that the projection matrices are orthogonal, and using the spectral resolution theorem we may write the representation matrix of **H** as

$$\mathsf{H} = \lambda_p D_p + \lambda_q D_q + \lambda_r D_r = \sum_{j=p,q,r} \lambda_j D_j \tag{10}$$

3. DERIVATION OF THE DENSITY MATRIX

Using eqs. (9) and (10), the mth power of **H** can be expressed by

$$\mathsf{H}^m = \lambda_p^m D_p + \lambda_q^m D_q + \lambda_r^m D_r = \sum_{i=p,q,r} \lambda_i^m D_i \tag{11}$$

When power m increases from 0 to 2 in eq. (11), the following simultaneous equations are defined:

$$\begin{aligned} E &= \lambda_p^0 D_p + \lambda_q^0 D_q + \lambda_r^0 D_r \\ \mathsf{H} &= \lambda_p D_p + \lambda_q D_q + \lambda_r D_r \\ \mathsf{H}^2 &= \lambda_p^2 D_r + \lambda_q^2 D_q + \lambda_r^2 D_r \end{aligned} \tag{12}$$

The terms on the left-hand side of these equations correspond to the power of the Hamiltonian operators. The right-hand side of (11) corresponds to the progression of $\lambda_i^m D_i \; (i = p.q.r)$. For example, if the system consists of six eigenvalues, density matrices can be obtained using six equations. The simultaneous equations that appear in eq. (12) are written concisely as the product of the vector and matrix:

$$\mathbf{H}_g = \Lambda_g D_g \tag{13}$$

where Λ_g is the Vandermonde matrix and $\mathbf{H}_p$ and D_g are the vectors defined by the following matrix representation:

$$\Lambda_g = \begin{bmatrix} \lambda_p^0 & \lambda_q^0 & \lambda_r^0 \\ \lambda_p & \lambda_q & \lambda_r \\ \lambda_p^2 & \lambda_q^2 & \lambda_r^2 \end{bmatrix}, \ \mathbf{H}_g = \begin{bmatrix} E \\ \mathbf{H} \\ \mathbf{H}^2 \end{bmatrix}, \ D_g = \begin{bmatrix} D_p \\ D_q \\ D_r \end{bmatrix} \tag{14}$$

The vector of the density matrices, D_g, can be calculated through the inverse of Vandermonde matrix Λ_g^{-1}, that is,

$$D_g = \Lambda_g^{-1} \mathbf{H}_g \tag{15}$$

where we write the i,jth element of a minor matrix of Λ_g as $(\Lambda_g)_{ij}$, the following equations are obtained for the density matrices:

$$D_p = \tfrac{1}{|\Lambda_g|}\left((\Lambda_g)_{11}E + (\Lambda_g)_{21}\mathbf{H} + (\Lambda_g)_{31}\mathbf{H}^2\right)$$

$$D_q = \tfrac{1}{|\Lambda_g|}\left((\Lambda_g)_{12}E + (\Lambda_g)_{22}\mathbf{H} + (\Lambda_g)_{32}\mathbf{H}^2\right) \tag{16}$$

$$D_r = \tfrac{1}{|\Lambda_g|}\left((\Lambda_g)_{13}E + (\Lambda_g)_{23}\mathbf{H} + (\Lambda_g)_{33}\mathbf{H}^2\right)$$

where $|\Lambda_g|$ is the determinant of matrix Λ_g. Thus, the density matrix corresponds to each eigenvalue whose values are known and nondegenerate, and it can be calculated when representation Hamiltonian H_g is definitely obtained.

The reason for using λ_p^0, λ_q^0, and λ_r^0 in eqs. (14) and (16) is due to the fact that if one of eigenvalues λ_p, λ_q, or λ_r is zero, then the numerical value of λ_u^0 becomes 1, so only the 0th power term of an eigenvalue will survive, and other terms of the higher power become zero.

4. THE EXPONENTIAL FORM OF THE ANGULAR MOMENTUM

An exponential of the angular momentum is expanded as a Taylor Series as follows:

$$\exp(\mathbf{H}) = \mathbf{H}^0 + \mathbf{H}^1 + \tfrac{1}{2!}\mathbf{H}^2 + \tfrac{1}{3!}\mathbf{H}^3 \cdots \tag{17}$$

Using eq. (11), this expansion becomes

$$\exp(\mathbf{H}) = \left(\lambda_p^0 + \lambda_p + \tfrac{1}{2!}\lambda_p^2 + \tfrac{1}{3!}\lambda_p^3 + \cdots\right)D_p$$

$$+ \left(\lambda_q^0 + \lambda_q + \tfrac{1}{2!}\lambda_q^2 + \tfrac{1}{3!}\lambda_q^3 + \cdots\right)D_q \tag{18}$$

$$+ \left(\lambda_r^0 + \lambda_r + \tfrac{1}{2!}\lambda_r^2 + \tfrac{1}{3!}\lambda_r^3 + \cdots\right)D_r$$

and so the exponential form of **H** can be obtained via an algebraic equation:

$$
\begin{aligned}
\exp(\mathbf{H}) &= \exp(\lambda_p)D_p + \exp(\lambda_q)D_q + \exp(\lambda_r)D_r \\
&= \sum_{j=p,q,r} \exp(\lambda_j)D_j
\end{aligned}
\tag{19}
$$

Likewise, the exponential of –**H** is given by

$$
\exp(-\mathbf{H}) = \sum_{j=p,q,r} \exp(-\lambda_j)D_j
\tag{20}
$$

5. ANGULAR MOMENTUM IN CARTESIAN COORDINATES

The angular momentum has three components — M_x, M_y, and M_z — along the x-, y-, and z-axes in a Cartesian coordinate system. These components satisfy the eigenvalue–eigenfunction equation, with eigenvalue matrix, Λ_g :

$$
\Lambda_g = \begin{bmatrix} \lambda_p & 0 & 0 \\ 0 & \lambda_q & 0 \\ 0 & 0 & \lambda_r \end{bmatrix}
\tag{21}
$$

When an angular momentum is directed toward arbitrary directions that have direction cosines $\cos\alpha$, $\cos\beta$, and $\cos\gamma$, where α, β, and γ are Euler angles, angular momentum $M(\alpha,\beta,\gamma)$ may be written as follows:

$$
M(\alpha,\beta,\gamma) = M_x \cos\alpha + M_y \cos\beta + M_z \cos\gamma
\tag{22}
$$

and therefore the following relation exists:

$$
\cos^2\alpha + \cos^2\beta + \cos^2\gamma = 1
\tag{23}
$$

Components M_x, M_y, and M_z of the angular momentum may be regarded as a projection onto the orthogonal coordinates. the eigen-equation is then written

$$
M(\alpha,\beta,\gamma)\cdot\phi(\alpha,\beta,\gamma) = \phi(\alpha,\beta,\gamma)\cdot\Lambda_M
\tag{24}
$$

We have assumed that $M(\alpha,\beta,\gamma)$, $\phi(\alpha,\beta,\gamma)$, and $\phi^{-1}(\alpha,\beta,\gamma)$ are 3×3 square matrices, and so the following relationship exists:

$$
\phi(\alpha,\beta,\gamma)\cdot\phi^{-1}(\alpha,\beta,\gamma) = E
\tag{25}
$$

Matrix $M(\alpha,\beta,\gamma)$ can be resolved according to eq. (9) such that

$$M(\alpha,\beta,\gamma) = \phi(\alpha,\beta,\gamma) \cdot \Lambda_M \cdot \phi^{-1}(\alpha,\beta,\gamma)$$
$$= \sum_{u=a,b,c} \lambda_u \phi(\alpha,\beta,\gamma) \cdot E_u \cdot \phi^{-1}(\alpha,\beta,\gamma) \tag{26}$$
$$= \sum_{u=a,b,c} \lambda_u P(\alpha,\beta,\gamma)_u$$

where

$$P(\alpha,\beta,\gamma)_u = \phi(\alpha,\beta,\gamma) \cdot E_u \cdot \phi^{-1}(\alpha,\beta,\gamma), \quad u = a, b, c \tag{27}$$

Eigenfunction matrix $\phi(\alpha,\beta,\gamma)$ and its inverse, $\phi^{-1}(\alpha,\beta,\gamma)$, are obviously dependent on the Euler angles. As a result, projection matrices $P(\alpha,\beta,\gamma)_u$ are also dependent on the direction.

Note that when one of the direction cosines, for example, $\cos\gamma$, is unity in eq. (23), the remaining terms $\cos^2\alpha$ and $\cos^2\beta$ should be simultaneously zero. The following relations must therefore be introduced for the sake of removing infinite values:

$$\cos\alpha + i\cos\beta = 1, \quad \cos\alpha - i\cos\beta = 0 \tag{28}$$

6. INTRINSIC DENSITY MATRIX OF ANGULAR MOMENTUM

Direction-dependent matrices $P(\alpha,\beta,\gamma)_u$ might be transformed supposing there exist matrices $T(\alpha,\beta,\gamma)$ and $T^{-1}(\alpha,\beta,\gamma)$ such that

$$D_u = T(\alpha,\beta,\gamma)^{-1} \cdot P(\alpha,\beta,\gamma) \cdot T(\alpha,\beta,\gamma)$$
$$= T(\alpha,\beta,\gamma)^{-1} \cdot \phi(\alpha,\beta,\gamma) \cdot E_u \cdot \phi(\alpha,\beta,\gamma)^{-1} \cdot T(\alpha,\beta,\gamma) \tag{29}$$

where $u=a$, b, and c, and

$$\phi(\alpha,\beta,\gamma) = T(\alpha,\beta,\gamma) \cdot \phi_H$$
$$\phi(\alpha,\beta,\gamma)^{-1} = \phi_H^{-1} \cdot T(\alpha,\beta,\gamma)^{-1} \tag{30}$$

Eigenfunction matrix ϕ_H can be written as

$$\phi_H = \begin{bmatrix} a_a & b_a & c_a \\ a_b & b_b & c_b \\ a_c & b_c & c_c \end{bmatrix} \tag{31}$$

where characters a, b, and c connect a basis set. The corresponding inverse of the eigenfunction matrix is written as

$$\phi_H^{-1} = \begin{bmatrix} a_a^* & b_a^* & c_a^* \\ a_b^* & b_b^* & c_b^* \\ a_c^* & b_c^* & c_c^* \end{bmatrix} \tag{32}$$

with the result that

$$\phi_H \cdot \phi_H^{-1} = E \tag{33}$$

and so

$$D_u = \phi_H \cdot E_u \cdot \phi_H^{-1} \tag{34}$$

Finally, matrix D_u ($u=a$, b, and c) is defined as the Intrinsic Density Matrix, which is independent of direction:

$$P(\alpha,\beta,\gamma)_u = T(\alpha,\beta,\gamma)^{-1} \cdot D_u \cdot T(\alpha,\beta,\gamma)$$
$$= T(\alpha,\beta,\gamma)^{-1} \cdot \phi(\alpha,\beta,\gamma) \cdot E_u \cdot \phi(\alpha,\beta,\gamma)^{-1} \cdot T(\alpha,\beta,\gamma) \tag{35}$$

Furthermore, $T(\alpha,\beta,\gamma)$ and $T(\alpha,\beta,\gamma)^{-1}$ are resolved in the following formula:

$$T(\alpha,\beta,\gamma) = T(\alpha,\beta) \cdot T(\gamma) \cdot S \cdot T(\gamma)^{-1} \tag{36}$$

and

$$T(\alpha,\beta,\gamma)^{-1} = T(\gamma) \cdot S \cdot T(\gamma)^{-1} \cdot T(\alpha,\beta)^{-1} \tag{37}$$

where matrices $T(\alpha,\beta)$ and $T(\gamma)$ and their inverses are given by the following:

$$T(\alpha,\beta) = \begin{bmatrix} (\cos[\alpha] - I\cos[\beta])^{-2} & 0 & 0 \\ 0 & (\cos[\alpha] - I\cos[\beta])^{-1} & 0 \\ 0 & 0 & 1 \end{bmatrix}$$

$$T(\gamma) = \begin{bmatrix} \cos[\gamma]^2 & 0 & 0 \\ 0 & \cos[\gamma] & 0 \\ 0 & 0 & 1 \end{bmatrix}$$

$$T(\alpha,\beta)^{-1} = \begin{bmatrix} (\cos[\alpha] + I\cos[\beta])^2 & 0 & 0 \\ 0 & (\cos[\alpha] + I\cos[\beta]) & 0 \\ 0 & 0 & 1 \end{bmatrix}$$

$$T(\gamma)^{-1} = \begin{bmatrix} \cos[\gamma]^{-2} & 0 & 0 \\ 0 & \cos[\gamma]^{-1} & 0 \\ 0 & 0 & 1 \end{bmatrix} \tag{38}$$

Matrices $T(\alpha,\beta)$ and $T(\gamma)$ and their inverses are obtained in the case of spin quantum number 1, but in the case of higher multiplicities the corresponding matrices are patterned after those of spin quantum number 1. Matrices S and S^{-1} are triangular and independent of direction; however, for $I > 1$ the individual spin matrices must be calculated:

$$S = \begin{bmatrix} 1 & \sqrt{2} & 1 \\ 0 & 1 & \sqrt{2} \\ 0 & 0 & 1 \end{bmatrix}$$

$$S^{-1} = \begin{bmatrix} 1 & -\sqrt{2} & 1 \\ 0 & 1 & -\sqrt{2} \\ 0 & 0 & 1 \end{bmatrix} \tag{39}$$

Thus, the eigenfunction matrices obtained from angular momentum $M(\alpha,\beta,\gamma)$ can be written as

$$\phi(\alpha,\beta,\gamma) = T(\alpha,\beta) \cdot T(\alpha,\beta) \cdot T(\gamma) \cdot S \cdot T(\gamma)^{-1} \cdot \phi_H \tag{40}$$

and

$$\phi(\alpha,\beta,\gamma)^{-1} = \phi_H^{-1} \cdot T(\gamma) \cdot S \cdot T(\gamma)^{-1} T \cdot (\alpha,\beta)^{-1} \tag{41}$$

From this result an intrinsic eigenfunction matrix is known to be equal to an eigenfunction matrix of M_H. Finally, angular momentum $M(\alpha,\beta,\gamma)$, having direction cosines $\cos\alpha$, $\cos\beta$, and $\cos\gamma$, can be written as

$$M(\alpha,\beta,\gamma) = \sum_{u=a,b,c} \left(\lambda_u T(\alpha,\beta) \cdot T(\gamma) \cdot S \cdot T(\gamma)^{-1} \right)$$

$$\cdot \phi_H \cdot E_u \cdot \phi_H^{-1} \cdot T(\gamma) \cdot S \cdot T(\gamma)^{-1} \cdot T(\alpha,\beta)^{-1} \tag{42}$$

When the components of $M(\alpha,\beta,\gamma)$ fall along an orthogonal coordinate, substitution of α, β, and γ must be made. This is especially so for the case of M_H, in which it is necessary to make substitutions using the transformation matrices.

7. COMMUTATION RELATIONS OF THE ANGULAR MOMENTA

Suppose there exists a second angular momentum designated as $M(\alpha',\beta',\gamma')$ with the same eigenvalue matrix, Λ_H :

$$M(\alpha',\beta',\gamma') = \sum_{u=a,b,c} \left(\lambda_u T(\alpha',\beta',\gamma') \cdot D_u \cdot T(\alpha',\beta',\gamma')^{-1} \right) \tag{43}$$

Then, of course, the following relation exists between the direction cosines:

$$\cos^2\alpha' + \cos^2\beta' + \cos^2\gamma' = 1 \tag{44}$$

and, supposing the angle between the two angular momentum vectors is θ,

$$\cos\alpha\cos\alpha' + \cos\beta\cos\beta' + \cos\gamma\cos\gamma' = \cos\theta \tag{45}$$

An exchange relation between $M(\alpha,\beta,\gamma)$ and $M(\alpha',\beta',\gamma')$ can be written as

$$M(\alpha,\beta,\gamma) \cdot M(\alpha',\beta',\gamma') = \sum_{u=a,b,c} \lambda_u T(\alpha,\beta,\gamma) \cdot D_u \cdot T(\alpha,\beta,\gamma)^{-1}$$

$$\cdot \sum_{v=a,b,c} \lambda_v T(\alpha',\beta',\gamma') \cdot D_v \cdot T(\alpha',\beta',\gamma')^{-1} \tag{46}$$

$$= \sum_{u=a,b,c} \sum_{v=a,b,c} \left(\lambda_u \lambda_v T(\alpha,\beta,\gamma) \cdot D_u \cdot T(\alpha,\beta,\gamma)^{-1} \cdot T(\alpha',\beta',\gamma') \cdot D_v \cdot T(\alpha',\beta',\gamma')^{-1} \right)$$

$$M(\alpha',\beta',\gamma') \cdot M(\alpha,\beta,\gamma) = \sum_{v=a,b,c} \lambda_v T(\alpha',\beta',\gamma') \cdot D_v \cdot T(\alpha',\beta',\gamma')^{-1}$$

$$\cdot \sum_{v=a,b,c} \lambda_u T(\alpha,\beta,\gamma) \cdot D_v \cdot T(\alpha,\beta,\gamma)^{-1} \tag{47}$$

$$= \sum_{v=a,b,c} \sum_{u=a,b,c} \left(\lambda_v \lambda_u T(\alpha',\beta',\gamma') \cdot D_v \cdot T(\alpha',\beta',\gamma')^{-1} \cdot T(\alpha,\beta,\gamma) \cdot D_u \cdot T(\alpha,\beta,\gamma)^{-1} \right)$$

As a product of transformation matrices $T(\alpha,\beta,\gamma)^{-1} \cdot T(\alpha',\beta',\gamma')$ and $T(\alpha',\beta',\gamma')^{-1} \cdot T(\alpha,\beta,\gamma)$, the result is a set of scalar quantities that corresponds to the u,v- and v,u- components of the matrices:

$$(\phi_H^{-1} \cdot T(\alpha,\beta,\gamma)^{-1} \cdot T(\alpha',\beta',\gamma') \cdot \phi_H)_{uv}$$

$$(\phi_H^{-1} \cdot T(\alpha',\beta',\gamma')^{-1} \cdot T(\alpha,\beta,\gamma) \cdot \phi_H)_{vu} \tag{48}$$

8. ANGULAR MOMENTUM UNDER THE HAMILTONIAN

When angular momentum $M(\alpha\beta\gamma)$ is governed by Hamiltonian H_0, this is transformed with a rotation operator to $M_1(\alpha\beta\gamma)$, such that

$$M_1(\alpha\beta\gamma) = \exp(-i\mathbf{H}_0 t)\cdot M_0(\alpha\beta\gamma)\cdot\exp(i\mathbf{H}_0 t) \tag{49}$$

where $\exp(-i\mathbf{H}_0 t)$ and $\exp(i\mathbf{H}_0 t)$ are a rotation operator of the Hamiltonian and its complex conjugate, respectively.

When the Hamiltonian is defined as the Zeeman interaction between an angular momentum with direction cosines $\cos\xi_0$, $\cos\eta_0$, and $\cos\varsigma_0$, and magnetic field $H_0(\xi_0\eta_0\varsigma_0)$, which is aligned, a rotation operator is expressed, assuming gyromagnetic ratio γ_m, as

$$\mathbf{H}_0 = \gamma_m H_0(\xi_0\eta_0\varsigma_0)\cdot M_0(\xi_0\eta_0\varsigma_0) = \omega_0 M_0(\xi_0\eta_0\varsigma_0) \tag{50}$$

where $\omega_0 = \gamma_m H_0$. The eigenvalues are obtained as $\Lambda_\mathbf{H}$, whose diagonal matrices are

$$\Lambda_\mathbf{H} = \omega_0 \begin{bmatrix} \lambda_p & 0 & 0 \\ 0 & \lambda_q & 0 \\ 0 & 0 & \lambda_r \end{bmatrix} \tag{51}$$

The rotation operator of $\mathbf{H}_0$ is given as

$$\begin{aligned}
\exp(i\omega_0 M_0(\xi_0\eta_0\gamma_0)t) &= \exp(i\lambda_p\omega_0 t)P_0(\xi_0\eta_0\gamma_0)_p \\
&\quad - \exp(i\lambda_q\omega_0 t)P_0(\xi_0\eta_0\gamma_0)_q - \exp(i\lambda_r\omega_0 t)P_0(\xi_0\eta_0\gamma_0)_r \\
&= \sum_{j=p,q,r} \exp(i\lambda_j\omega_0 t)P_0(\xi_0\eta_0\gamma_0)_j
\end{aligned} \tag{52}$$

and its complex conjugate is written as

$$\exp(-i\omega_0 M_0(\xi_0\eta_0\gamma_0)t) = \sum_{j=p,q,r} \exp(-i\lambda_j\omega_0 t)P_0(\xi_0\eta_0\gamma_0)_j \tag{53}$$

The resultant angular momentum that is transformed by the Zeeman interaction is given by

$$\begin{aligned}
M_1(\alpha\beta\gamma) &= \exp(i\omega_0 M_0(\xi_0\eta_0\gamma_0)t)\cdot M_0 \cdot \exp(-i\omega_0 M_0(\xi_0\eta_0\gamma_0)t) \\
&= \sum_{j=p,q,r}\sum_{k=p,q,r}\left[\lambda_u\exp(-i(\lambda_j-\lambda_k)\omega_0 t)\cdot P_0(\xi_0\eta_0\gamma_0)_j\cdot M_0(\alpha\beta\gamma)\cdot P_0(\xi_0\eta_0\gamma_0)_k\right] \\
&= \sum_{j=p,q,r}\sum_{u=p,q,r}\sum_{k=p,q,r}\begin{bmatrix}\lambda_u\exp(-i(\lambda_j-\lambda_k)\omega_0 t)\cdot T_0(\xi_0\eta_0\gamma_0)\cdot\phi_H\cdot E_j\cdot\phi_H^{-1}\cdot T_0(\xi_0\eta_0\gamma_0)^{-1}\cdot \\ T_0(\xi_0\eta_0\gamma_0)\cdot\phi_H\cdot E_u\cdot\phi_H^{-1}\cdot T_0(\xi_0\eta_0\gamma_0)^{-1}\cdot \\ T_0(\xi_0\eta_0\gamma_0)\cdot\phi_H\cdot E_k\cdot\phi_H^{-1}\cdot T_0(\xi_0\eta_0\gamma_0)^{-1}\end{bmatrix}
\end{aligned} \tag{54}$$

Here we take the term between E_j and E_u, and that between E_u and E_k in eq. (54):

$$E_j \cdot \phi_\mathsf{H}^{-1} \cdot T_0(\xi_0 \eta_0 \varsigma_0)^{-1} \cdot \phi(\alpha\beta\gamma) \cdot \phi_\mathsf{H} \cdot E_u$$
$$E_u \cdot \phi_\mathsf{H}^{-1} \cdot T_0(\xi_0 \eta_0 \varsigma_0)^{-1} \cdot \phi(\alpha\beta\gamma) \cdot \phi_\mathsf{H} \cdot E_k \tag{55}$$

The product of these scalar quantities decides a transition between states λ_j and λ_k.

9. SUCCESSIVE TRANSFORMATIONS

Under any kind of second Zeeman interaction, $\mathbf{H}_1$, including angular momentum $M_1(\xi_1 \eta_1 \varsigma_1)$, angular momentum $M_1(\alpha\beta\gamma)$ is transformed so as to obtain $M_2(\alpha\beta\gamma)$:

$$M_2(\alpha\beta\gamma) = \exp(-i\mathbf{H}_1 t) \cdot M_1(\alpha\beta\gamma) \cdot \exp(i\mathbf{H}_1 t) \tag{57}$$

The Hamiltonian is given as the Zeeman interaction between oscillating field $H_1 \cos\omega t$ and angular momentum $M_1(\xi_1 \eta_1 \varsigma_1)$, such as

$$H_1 = \gamma_m H_1 \cos\omega_0 t \cdot M(\xi_1 \eta_1 \varsigma_1)$$
$$= \omega_1 \cos\omega_0 t \cdot M(\xi_1 \eta_1 \varsigma_1) \tag{58}$$

Thus

$$M_2(\alpha\beta\gamma) = \exp(i\omega_1 M_1(\xi_1 \eta_1 \gamma_1)t) \cdot M_1(\alpha\beta\gamma) \cdot \exp(-i\omega_1 M_1(\xi_1 \eta_1 \gamma_1)t)$$
$$= \sum_{m=p,q,r} \exp(-i\lambda_{m-1} t) P_1(\xi_1 \eta_1 \gamma_1)_j \cdot M_1(\alpha\beta\gamma) \cdot \sum_{n=a,b,c} \exp(i\lambda_n \omega_1 t) P_1(\xi_1 \eta_1 \gamma_1)_k$$
$$= \sum_{j=p,q,r} \sum_{k=p,q,r} \left[\lambda_u \exp(-i(\lambda_j - \lambda_k)\omega_1 t) \cdot P_1(\xi_1 \eta_1 \gamma_1)_j \cdot M_1(\alpha\beta\gamma) \cdot P_1(\xi_1 \eta_1 \gamma_1)_k \right]$$
$$= \sum_{j=p,q,r} \sum_{u=p,q,r} \sum_{k=p,q,r} \begin{bmatrix} \lambda_u \exp(-i(\lambda_j - \lambda_k)\omega_1 t) \cdot T_1(\xi_1 \eta_1 \gamma_1) \cdot \phi_H \cdot E_j \cdot \phi_H^{-1} \cdot T_1(\xi_1 \eta_1 \gamma_1)^{-1} \cdot \\ T_1(\xi_1 \eta_1 \gamma_1) \cdot \phi_H \cdot E_u \cdot \phi_H^{-1} \cdot T_1(\xi_1 \eta_1 \gamma_1)^{-1} \cdot \\ T_1(\xi_1 \eta_1 \gamma_1) \cdot \phi_H \cdot E_k \cdot \phi_H^{-1} \cdot T_1(\xi_1 \eta_1 \gamma_1)^{-1} \end{bmatrix} \tag{59}$$

The exponential terms of eqs. (54) and (59) are extracted and combined in order to define the oscillatory behavior of $M_1(\xi_1 \eta_1 \varsigma_1)$, that is,

$$\sum_{j=p,q,r} \sum_{k=p,q,r} \left[\exp(-i(\lambda_j - \lambda_k)\omega_1 t) \sum_{m=p,q,r} \sum_{n=p,q,r} \exp(-i(\lambda_m - \lambda_n)\omega_0 t) \right]$$
$$= \sum_{j=p.q.r} \sum_{k=p,q,r} \sum_{m=p,q,r} \sum_{n=p,q,r} \exp(-i(\lambda_j - \lambda_k)\omega_1 - (\lambda_m - \lambda_n)\omega_0)t) \tag{60}$$

This term is very informative with respect to detection of a signal in the time domain. Usually a single harmonic of frequency ω_0 is used so that the first harmonic of ω_1 shows a single quantum transition and the second double quantum.

10. DISCUSSION

The density matrix formalism of angular momentum with half- and integer-spin quantum numbers has been presented using the spectrum dissolving theorem, which yields an exact transformation when the rotation operator is used. Next, the spectral resolution theorem enabled obtaining a matrix representation of the Hamiltonian as a linear combination of the matrix multiplied by its associated non-degenerate eigenvalue. Each matrix is a projection onto a basis function and orthogonal to other members of the set. The matrix also has the property that its square returns the same matrix, and therefore the nth power of a matrix is a linear combination of the projection matrix multiplied by the nth power of the associated eigenvalue. This fact leads to an algebraic expression of an exponential of the Hamiltonian. This procedure of the projection onto the basis might be topological.

Treatment of angular momentum in a multi-quantum regime naturally benefits from this operation. We applied this theorem to quantum operator $M(\alpha\beta\gamma)$, defined as $M_x \cos\alpha + M_y \cos\beta + M_z \cos\gamma$, in which $\cos\alpha$, $\cos\beta$, and $\cos\gamma$ are direction cosines of Euler angles α, β, and γ. The resultant eigenvalues of the angular momentum are independent of direction, but the projection matrices are dependent. The matrix for an angular momentum can therefore be resolved as a product of the transformation matrix with a direction cosine and an eigenfunction matrix that corresponds to the transition. The projection matrix, after removal of information on direction, is an intrinsic density matrix. The procedure to obtain M_x, M_y, or M_z from the operator entails a geometric projection onto an orthogonal coordinate. An exchange relation among elements M_x, M_y, and M_z of an angular momentum was then examined and used to generate transformation matrices that produce the projection matrix independent of the density matrix.

The vector composed of the density matrices is generated by taking the product of the inverse of Vandermonde matrix composed by the $(n-1)$th powers of the eigenvalues and the vector composed of the $(n-1)$th powers of each angular momentum, both including the 0th power. This procedure entails an interpolation problem of finding the polynomial and degree $n - 1$. When the eigenvalues are nondegenerate, all the density matrices are easily calculated.

Since a linear combination of the spin density matrix-associated eigenvalue directly yields a solution for the stochastic Liouville equation, one may evaluate the spin response under full or partial excitation. Consequently, the present formalism is applicable to the treatment of orbital angular momentum.

CONTENTS OF PREVIOUS VOLUMES

INDEX